The U.S. Paper Industry and Sustainable Production

An Argument for Restructuring

Maureen Smith

The MIT Press
Cambridge, Massachusetts
London, England

This book was set in Sabon on the Corel Ventura by Wellington Graphics.

Printed and bound in the United States of America.

The paper used in this publication is both acid and totally chlorine free (TCF). It meets the minimum requirements of American Standard for Information Sciences—Permanence of Paper for Printed Library Materials, ANSI Z39.48-1984.

Library of Congress Cataloging-in-Publication Data

Smith, Maureen, 1956–
The U.S. paper industry and sustainable production : an argument for restructuring / Maureen Smith.
p. cm. — (Urban and industrial environments)
Includes bibliographical references and index.
ISBN 0-262-19377-9 (hc : alk. paper)
1. Paper industry—Government policy—United States. 2. Wood-pulp industry—Government policy—United States. I. Title. II. Series.
HD9826.S65 1997
338.4'5676'0973—DC20 96-41991
CIP

In memory of my father
Tremaine Francis Smith

Contents

Series Foreword

Robert Gottlieb

During the past several decades, environmental discourse has significantly broadened to include perspectives that connect our understanding of human and natural environments. Environmental questions are increasingly seen as embedded in issues of technology, landscape, and social organization. Critical investigations of these questions have begun to be located in various arenas of interdisciplinary research that bring together the insights of such fields as industrial ecology, urban studies, industrial geography, urban environmental history, regional planning, and occupational and environmental health sciences. New ways of looking at public policy parallel these changes in environmental research and teaching within the university. The new approaches emphasize environmental justice, toxics use reduction, place-centered environmental management, and pollution prevention. Community and industry initiatives in such areas as land use, produce design and procurement, and resource management have also contributed new insights about urban and industrial environmental change. Urban and Industrial Environments, a new book series from The MIT Press, has been created to advance these concepts of urban and industrial environmental analysis. Books in this series are being drawn from the wide range of disciplines that address the environment in the context of daily life experiences and the evolving structures of industrial activity and urban form. The series offers views of the complex and contested arena of environmental policymaking, the role of social movements and interest groups in the formation of policy and changes in institutions, and the evolution of industry sectors.

Series Foreword

Acknowledgments

This book developed out of a project I began during my studies at the UCLA Graduate School of Architecture and Urban Planning in the late 1980s. The topic was sparked by a suggestion made by Robert Gottlieb, and the ensuing effort would not have been completed without his support and encouragement. I am deeply indebted for his help and his unwavering optimism; but most of all, I am inspired by the wonderful example he sets in his own work. In addition to Robert Gottlieb, Stephanie Pincetl provided thoughtful comments on several early drafts of the manuscript as well as strong encouragement to complete it. I also benefited at different stages in the process from comments provided by Louis Blumberg, Norm Masters, Michael Picker, Peter Sinsheimer, Al Wong, and some six anonymous reviewers. I am particularly grateful to Norm Masters for the many hours he spent on the phone with me discussing the MacMillan Bloedel project in West Sacramento, and for inviting me to be involved in his community.

My access to various individuals and forums benefited in important ways from the support of Randy Hayes, Susan Bluestone, and Adam Davis. The many interesting conversations I had that flowed from public presentations of parts of the manuscript and from my work with environmental groups in California were integral to the development of several chapters. In addition, the excellent research papers and other publications freely distributed by the USDA Forest Products Lab, and the collection at the UC Berkeley Forest Products Library made the challenging task of writing about the paper industry from southern California more manageable.

Although I discovered that books are never really completed and that writing them is indeed a lonely business, I also came to appreciate the invaluable role that friends and family play in the process. Susan and Dan Todd and my other river friends provided wonderful companionship (and dependable duct tape reserves) in wild places where I could both forget about the book and remember why it was worth writing in the first place. Marty Fujita, Dana Livingston, Susan Todd, and my sister, Frances Smith, were my sounding boards (sometimes at very long distance) for the low points. My deepest gratitude is reserved for my mother, Phyllis Smith. She informed me at the outset that I was crazy to take this on and then, in her own uniquely gracious style, provided unqualified support for the duration.

Introduction

The recycling movement of the 1980s had two distinctive characteristics. On one hand, recycling had been propelled back onto the national environmental agenda by the garbage crisis that emerged in the 1980s and immortalized in images of the wandering garbage barge from New York. The underlying strength of the movement was largely rooted in the burgeoning community-based "Not In My Backyard" opposition to toxic hazards that both the development of new landfill capacity and the construction of municipal solid waste incinerators would bring, and in the standoffs resulting when cities and counties attempted to site new facilities in the face of organized citizen resistance. Thus in this latest incarnation, recycling was almost exclusively framed as a solid waste management alternative and as such reflected a strikingly narrower set of concerns than had earlier recycling movements.

On the other hand, recycling had crystallized as a popular idea like never before. In the media events leading up to Earth Day 1990—an extravagant 20th anniversary celebration of the first Earth Day—recycling was undeniably at center stage. Reflecting a central theme of Earth Day events, which were strongly characterized by a focus on personal responsibility for the environment, recycling was portrayed as the single most accessible expression of the concept of "think globally, act locally": a way to square one's personal behavior with one's principles, and something everyone could do. "Recycling," observed one participant at the 1990 National Recycling Congress, "has a lot of power right now." It had become a mainstream movement supported by all the major environmental groups. Its organized forms, such as the National Recycling Coalition and various state and regional associations, were dominated by public and private solid waste professionals, independent recyclers who operated collection and brokering services, and various secondary materials industries and other businesses involved in recycling.

Although the federal government had largely abandoned municipal solid waste policy to the states in the early 1980s (nevertheless continuing to demonstrate a strong level of support for the development of solid waste incinerator projects), a number of public policy makers, ranging from city councils and county officials at the front lines of the solid waste crisis, to state legislators, were quick to join this immense groundswell of popular support for recycling. Although the policies and legislation that began to emerge around recycling and solid waste management by the mid-1980s were extremely heterogeneous and often haphazardly defined, they had reached flood-stage by the late 1980s. In 1988 alone, one organization estimated that approximately two thousand solid waste bills had been introduced in state legislatures.[1]

In the midst of celebrating and legislating this powerful new commitment to recycling, however, some problems began to emerge, especially around wastepaper. The single largest component of municipal solid waste, wastepaper had been a major focus of the new recycling legislation, and old newsprint (ONP) in particular had been among the first materials to be targeted (along with aluminum cans and glass) in the rapidly growing number of mandatory collection programs. In late 1988, and increasingly throughout 1989 and 1990, articles and editorials began to appear with headlines such as "Surfeit of Used Newsprint Erases Profit in Recycling," "Some Newspaper is Landfilled; Programs Halt Collection Efforts," and, succinctly, "Is Recycled Paper a Waste?"[2] With, apparently, good intentions all around, it nevertheless appeared that for ONP at least, the market had collapsed.

Around the same time, although at a less publicly visible level, tentative questions began to be raised regarding the potentially negative environmental impacts of secondary materials-based manufacturing processes themselves. In the case of paper recycling, these early concerns were expressed primarily in relation to the toxicity of process sludge contaminated by heavy metals from inks and by PCBs, which were thought to come from the recycling of older carbonless copy papers. They largely served to introduce the topic of chemical pollution as a serious element of the modern recycling dialogue among environmental groups and solid waste professionals. However, these concerns also intersected with a sharply rising level of concern over pollution from virgin woodpulp mills. Focused initially on the formation and release of dioxins and other highly toxic organochlorine compounds formed as by-products of chlorine-based pulp bleaching processes, these concerns eventually carried over into the subject of paper recycling technology in terms of the direct use of chlorine and chlorine compounds in deinking processes, as well as in the form of suggestions that deinking chlorine-bleached wastepaper might somehow concentrate the dioxin problem.

As the investigation of wastepaper processing technologies got underway, however, the growing controversy over organochlorine and other toxic chemical pollution from woodpulp mills was accelerated by the 1989 publication (based on 1987 data) of the first figures from the new Toxics Release Inventory (TRI)—a database of chemical releases compiled by the EPA from mandatory reports filed by large manufacturing facilities. Out of all the major manufacturing sectors, the paper industry ranked third (behind chemicals and primary metals manufacturers) in total releases of roughly three hundred reportable toxic chemicals, and second for releases to air, averaging hundreds of millions of pounds per facility of total annual releases of TRI chemicals.[3]

Although the late 1980s recycling movement was dominated by a solid waste framework, the exploding popular literature on recycling had in fact tended to include at least some claim to the more historically celebrated aspects of recycling, such as pollution reduction and resource conservation. During the oil crises of the 1970s, for example, energy conservation advantages had provided one of the strongest arguments for increased recycling (although ultimately with little effect). A few years earlier, as a contribution to the first Earth Day, Garrett de Bell had written, "Recycling is a major part of the solution of many environmental problems. It is important to air and water pollution and to wilderness preservation."[4]

For the most part, however, these other environmental arguments for recycling, almost totally unexplored since the 1970s, were simply tacked onto modern recycling rhetoric in the form of a kind of assumed "bonus package" of benefits. Yet, as a few analysts began to seek information not only to confirm some of these other advantages, but to explore the pollution implications of potentially substantially increased secondary materials-based manufacturing (manufacturing that was likely to be situated near population centers—"urban forests" or "urban mines"—that supplied the secondary materials, they found a major void in the technical and policy literature on recycling technology and its environmental impacts.

While the national recycling movement was approaching its crescendo in 1990, a different story was unfolding in the Pacific Northwest where decades of conflict over the decimation of what had come to be known as the "ancient forest" were coming to a full boil. Although the conflicts were longstanding, they had recently escalated to national proportions due to controversy over the prospective listing of the Northern Spotted Owl as an endangered species and to comparatively aggressive new initiatives such as California's "Big Green" and "Forests Forever" ballot propositions. Although ultimately defeated in the face of well-funded industry opposition, these propositions reflected an increasingly widespread public understanding of and opposition to the forest policies being played out not only in the western United States, but throughout North and South America, southeast Asia, and elsewhere in the world.

Ironically, although the connection between trees and paper is generally recognized by the public, it has long had a sort of mythic status—one rarely informed in terms of how much timber, from where, and for what products. The connection between trees and wastepaper has been even more obscure. As an environmental issue, the enormous impact that paper production has on forests, and the potential for it to be diminished by increased paper recycling, had no real constituency of its own. Despite the resuscitation of 1970s rhetoric about saving trees by recycling, forest conservation and forest advocates were not directly connected with the recycling movement at the time, and forest policy and solid waste policy remained poles apart. Nevertheless, forests were being wrecked, almost everyone was upset regardless of which side they were on, and paper recycling was supposed to save trees. Clearly, a connection needed to be clarified.

Against this background, in late 1989 and early 1990, the questions that eventually led to this book were originally formulated. Focused initially on paper recycling, this study sought to investigate three related questions. First, taking off from the collapse of the ONP market, what were paper manufacturers doing about increasing their capacity to utilize wastepaper, and were they being "sluggish"? Second, what did paper recycling really offer in the way of energy conservation and pollution prevention? Was it a "clean" technology, at least substantially cleaner than woodpulp production, or did it raise unique hazards? Finally, what exactly was the connection between paper recycling and forests? The objective was to develop not only a perspective on improved policy and legislative approaches to increased paper recycling, but also an understanding of increased secondary materials use that more accurately reflected its broader environmental and economic dimensions.

During the late 1980s and early 1990s, as recycling development took hold, and as the controversy over forest management and timber supply continued to escalate, both the chlorine issue and the general subject of pollution from virgin pulp production mushroomed in significance and began to drive the industry in equally powerful ways. At the same time the subject of tree-free paper, or the use of annual plant fibers (nonwoods) in papermaking, emerged on the national radar screen in the early 1990s. As they began to promote this compelling alternative fiber supply, advocates pointed to both decades of research by the USDA and the established use of annual fiber crops and agricultural wastes for papermaking in other countries. They suggested a range of potential benefits, which in several respects mirrored those thought to be associated with paper recycling.

As important as each of these issues is in its own right, what ultimately becomes clear is that the most significant underlying questions concern the relationship between them, and the degree to which the opportunities and barriers in each area are linked by the nature and structure of the pulp and paper industry itself. Situated within the context of the forest products industries as a whole, the paper industry is the nexus of a complex set of environmental relationships that run from the resource base to the postconsumer disposition of wastepaper. Although this may in some respects be obvious, an industry-centered perspective on materials utilization and environmental impact—a central theme explored in this book—in fact represents a significant departure from dominant frameworks in which environmental issues are usually approached. Few environmental policy initiatives have adequately represented either the primacy of the industrial role or the degree to which it connects and arbitrates a range of environmental issues that may seem remote and disparate to those who specialize in any one issue.

Recycling policy is illustrative. Recycling, as the word itself implies, has come to be understood as a circular process of materials use. It is frequently divided into three or more broad phases, generally: the recovery of recyclable materials, the use of recovered materials by manufacturers, and the purchase and use of products containing recovered materials. The contemporary recycling movement has focused primarily on the first phase, and more recently on the last, but relatively little on the critical industrial link between the two. This has been reflected in the pervasiveness of narrowly framed supply-and-demand approaches to recycling policy. Whether viewed in terms of barriers or incentives, supply-side approaches have typically focused on the operational and financial logistics of municipal recovery programs such as residential and commercial collection schemes, have debated the subject of extensive source separation versus intermixed collection with centralized materials separation facilities, and have examined mechanisms for brokering recovered materials. The theory, of course, is that when secondary materials supplies become plentiful, cheap, and dependable as a function of improved supply-side programs, manufacturers will naturally switch to them. Demand-side approaches attempt to create favorable market conditions for recycled products. They focus on government procurement programs and on other mandatory or voluntary standards for recycled content in products, or on "green consumption" in general.

In the debates that have been underway since the late 1980s over the reauthorization of the federal Resource Conservation and Recovery Act (RCRA), the primary federal law affecting municipal solid waste management and thus recycling, "markets" has become the key buzzword. It ambiguously refers to both intermediate manufacturing markets for recovered materials, and to commercial and end-user markets for recycled products. The manufacturer, situated between the recovered materials and consumers, is usually targeted, at best, as a sort of indeterminate middle market for secondary materials. This intermediary, it is hoped, is sensitive to at least some forms of remote control by manipulation of external supply-and-demand conditions, but has otherwise been treated as something of a black box.

Understood from an industry-centered perspective, however, recycling is fundamentally an issue of raw materials substitution or, at least, augmentation within a given industrial sector, with an array of associated technological and other implications. As such, the use of secondary materials, although clearly not unrelated to external supply and demand conditions for secondary materials and recycled products, has a powerful life and logic of its own that may be equally or more important in determining the level, form, and benefits of secondary materials use in manufacturing. This internal logic is highly dependent on such factors as the existing industrial infrastructure and capital commitment, materials supply lines, geographic distribution and location, dominant technologies, research and development capacity, economic integration with other industrial sectors, as well as internal and external structures and institutions that reflect both existing and historical forms of regulation and subsidy. Understanding and influencing the prospects not only for increased recycling, but also for the "cleanest" and most appropriate forms of secondary materials-based manufacturing, requires at a minimum a clear understanding of the industrial context in which potentially major shifts in basic materials, technology, and economic relationships—that is, industrial restructuring—must occur.

Among other things this view highlights the fact that we might expect very different circumstances and might need to develop very different sorts of approaches, depending on the materials we, as a society, are interested in recycling at accelerated rates. Such approaches might logically be expected to vary as widely as the key manufacturing and other industries that will make use of those materials do, ranging from primary metals manufacturing (aluminum, steel and other scrap metals) to oil refineries (used oil) to chemical and plastics industries (recovered plastic). The task, from this view, is less one of developing recycling industries, than it is one of *redeveloping* basic materials industries.

We must also bear in mind that as a society we have begun to demonstrate at least a rhetorical interest in imposing some discipline on our consumption of these materials as exemplified in the popular slogan "reduce, reuse, recycle" and in the position of "source reduction" at the top of officially adopted hierarchies of solid waste management priorities. One can hardly fail to recognize the significance of the mass production-oriented commodities producers in terms of prospects for pursuing a source reduction agenda.

In addition to the need to focus on the core industry itself, however, it further becomes evident that an approach to policies seeking to influence progress on solid waste reduction and other environmental goals by way of changing traditional patterns of materials use must deliberately accommodate a broad spectrum of associated environmental issues from the outset; it is not sufficient to assume that the "bonus package" of energy, pollution, and natural resource benefits will be automatically forthcoming. Thus, the second major theme of this study is integrated environmental analysis. Energy consumption and environmental pollution associated with pulp and paper production (whether based on virgin or secondary materials) are strongly interdependent and cannot be meaningfully considered in isolation from one another, just as downstream solid waste disposal problems cannot be isolated from natural resource problems nor from rising levels of per capita consumption.

In the absence of approaches based on integrative and systemic environmental analyses, expectations are often dashed, and important opportunities overlooked or foreclosed. At worst, serious distortions of intention can and do arise in strange and unpredictable forms and may be as likely to create new environmental problems as to solve those originally targeted. Many such instances have occurred in the recent history of the paper industry, instances that can be traced equally to traditional narrowly defined agendas and to the black box approach to industrial dynamics. Among those considered in the following chapters are the effects associated with the exaggerated emphasis on the use of postconsumer wastepaper as a panacea for paper industry environmental problems; a continuing and substantial disjuncture between forest conservation and wastepaper use (in effect, paper recycling without saving any trees); the strengthening of pollution control strategies that run counter to sustainable materials use policies; and early signs that the augmentation of the resource base that wastepaper has provided may have accelerated both total production and consumption of pulp and paper.

Those who have followed developments in pollution prevention policy may recognize the outlines of some familiar constructs here. Given national status by the passage of the federal Pollution Prevention Act of 1990 and the EPA's "pollution prevention strategy" outlined in 1991, pollution prevention has been widely embraced as the new paradigm that will guide the next generation of national efforts to reduce environmental pollution.

Although subject to conflicting and evolving interpretations, pollution prevention is usually broadly defined in contrast to the pollution control approach that has traditionally characterized toxic and hazardous substance regulation. The dominant characteristics of pollution control are expressed at the national level in a variety of key pieces of legislation ranging from the Clean Air and Clean Water Acts to the Occupational Safety and Health Act, RCRA, and others. They centrally incorporate (1) a direct-regulation (command-and-control) format, and (2) a medium-specific format (i.e., a focus on the individual "media" of environmental pollution, such as air, water, land, and consumer products). The inefficiencies of both the direct-regulation and medium-specific formats, it is proposed, will be overcome by a new, preventative paradigm that seeks to address multimedia pollution at "upstream" sources (such as at product and process design phases, and by reducing or eliminating the use of toxic substances) rather than at the "end of the pipe" (e.g., smokestack, effluent outfall pipe, landfill) where attention and resources have long been focused.

Much of the debate in defining pollution prevention has turned on which of these two aspects of pollution control is emphasized as the underlying take-off point for pollution prevention. On one hand, pollution prevention is said to be centrally defined by a new relationship between industry and government, in which interventionist and bureaucratic direct-regulation formats are to be forsaken (or at least partially supplanted) in favor of more voluntaristic and flexible methods that will strategically harness the power of competition to stimulate profitable clean technology and other environmentally beneficial innovations. On the other hand, the structural inadequacy of medium-specific regulation, evident in the historical sequence in which the key pollution control statutes emerged, provides the more important point of departure.

Whereas air and water pollution controls were the first to be developed, the regulation of industrial solid waste came later, partly because the problem of land disposal had become more serious as a consequence of air and water pollution control technologies that generated solid waste (e.g., by capturing smokestack particulates or dewatered sludges that would otherwise have been released in gaseous or liquid forms). The management of these solid or containable hazardous wastes, which relied heavily on landfills, incinerators, and underground injection wells, led full circle to new problems of toxic air pollution as well as ground and surface water contamination. This by-product of medium-specific regulation—in which limited relief in one domain is obtained in part by shifting an underlying problem to another domain—has much in common with recycling policy as defined and implemented in the past decade. Indeed, recycling policy, as an elaboration of solid waste management policy, is a progeny of the pollution control approach.

Pollution prevention thus provides one of several interesting and potentially useful frameworks within which to think about materials use, particularly in terms of what it has to say about the possibilities for different public policy/industry relations and for front-end or upstream preventative and integrative approaches. The subject will be considered at more length in the following chapters with particular attention to how an industrial focus may inform and guide both sets of issues.

The fit between evolving pollution prevention thinking and recycling or materials policy is presently somewhat uncertain, however, since not only have pollution control strategies traditionally revolved around separate media, but the overall environmental policy framework in which they are embedded has long reflected a strict separation between natural resource issues, industrial pollution control policy (and its stepchild, occupational health), and consumer and post-consumer issues. One need only consider the organization of federal agencies to comprehend the degree to which the isolation of these traditional policy areas has been institutionalized. Just as forest policy has been poles apart from municipal solid waste policy, so has industrial pollution control policy been isolated from them both.

Broader policy frameworks have begun to emerge, including those that attempt to understand both pollution prevention, and environmental policy overall, within a sustainable development context. Frances Irwin points out, "In a time when environmental issues are being reframed . . . the term *integration* appears in many guises." Integrated pollution control, as the first element of a pollution prevention approach, has been referred to as an *internal* integration of traditional single-medium approaches. However, it has also been referred to in conjunction with "*external* integration between environmental policy and other policy sectors." She notes, "[T]he assumption is that internal integration is a step toward external integration."[5] As has widely come to be recognized by both public and private sectors, the inevitable challenge these approaches ultimately face is that of integrating environmental policy with economic policy.

The term *sustainable development,* first popularized in *Our Common Future,* the 1987 report of the World Commission on Environment and Development, has come to broadly designate this new, more integrative paradigm.[6] It is often used somewhat interchangeably with terms such as *sustainable production, sustainable economies,* and *sustainable technologies* and has given rise to terms such as *sustainable cities.* Although each carries a slightly different focus, the prefix *sustainable* is generally based on the assertion that "the earth's natural resources have finite capabilities to support human production and consumption and that the continuation of existing economic policies risks irreversible damage to natural systems on which all life depends."[7] It has thus (although loosely and subject to highly contradictory interpretations) come to designate both some commitment to a deeper consideration of the complex relationships between environment and economy, and the need for greater equity within and between societies and generations.

Little attempt is made here to pin down a more critical, operational definition of the concept of sustainability. The term, as used in this book, is primarily intended to suggest the outlines of a larger, more long-term conceptual framework in which the more deliberate discussions of materials use, pollution prevention, and policy integration, and the analysis of a particular industrial sector may ultimately be embedded. In this context, the term *sustainable production* is favored over other variants to give emphasis to the need to focus on basic industries that in their structure and practice provide the most important linkages between the various social, economic, and ecological goals associated with sustainability.

The organization of this book is as follows. Chapters 1 and 2 provide necessary background by outlining the basic structural characteristics of the U.S. pulp and paper industry. The relationship of the industry to the larger forest products industry of which it is part, are considered, as well as its patterns of consumption of domestic and global fiber resources. Chapter 3 reviews the core technologies employed in virgin pulp and paper production with emphasis on their associated environmental impacts and on the role for technological innovation and comparative analysis. The use of chlorine and compounds in pulp production provides the basis for a short case study of pollution prevention in the context of the paper industry, and the use of nonwood fibers is considered in terms of a more expansive application of pollution prevention policy. The current status of domestic paper recycling with respect to both historical fiber utilization patterns and contemporary links to industrial development is considered in chapter 4. Chapter 5 presents a case study of a recent environmental review process for a proposed major paper recycling facility; this case study serves both to tie together elements of the preceding discussion and to highlight the local consequences of narrowly defined policies. Chapter 6 returns to the broader discussion of environmental policy developments and approaches that can be taken with respect to the industry. It raises the subject of different organizing vehicles for addressing the range of concerns associated with the prospects for influencing the course of a highly mature and highly integrated multinational industry, and it suggests the outlines of an agenda for pursuing this approach.

The twin themes of this book—of industry-centered and integrated environmental analysis—are fundamentally based on the dichotomy between environment and economy, but the object of study itself—the paper industry—provides a focal point around which we can consider both. If, indeed, it is necessary to understand the structure of the industry to influence its use of materials and reduce the burdens it places on the environment; and if the substantially increased use of alternative materials—driven by a range of environmental concerns—requires restructuring of the industry, it seems advisable that we have some idea of where we are heading and what we are taking on. The following pages are focused on the exploration of such questions.

1

An Overview of the U.S. Pulp and Paper Industry

In March of 1990, the leadership of the U.S. paper industry convened at the Waldorf-Astoria hotel in New York for the 113th annual "Paper Week" meetings. This event, which fell just a month before the climax of the Earth Day 1990 celebrations, also served to commemorate the 300th anniversary of the American paper industry, which dates its origins to a mill built by William Rittenhouse near Philadelphia in 1690. A good deal of attention at the meetings was focused on the weakening state of the economy, then dipping into recession. However, in a keynote speech Red Caveney, president of the American Paper Institute (at the time the industry's largest and most important trade association), took the opportunity to highlight the industry's acknowledgment of a leading environmental issue. He commented that: "the paper industry [has] renewed and bolstered its longstanding commitment to paper recycling in announcing a national goal of 40 percent wastepaper recovery for domestic recycling and export by the end of 1995."[1]

If it was somewhat paradoxical that the industry would express its commitment to paper recycling by way of announcing a national goal for wastepaper *recovery,* an effort that was already being led by the local and state governments responsible for establishing solid waste management programs, and that did not directly address the industry's plans for making use of this wastepaper, it was nevertheless true that the industry's use of secondary fiber sources had a long history. Indeed, the seminal Rittenhouse mill had relied exclusively on recovered fiber derived from cotton and linen rags and wastepaper.[2] In the context of the times, however, the wastepaper recovery goal that the industry set for the nation could be read as one way for it to simultaneously embrace and deflect a major environmental issue—an issue that had already begun to demonstrate its potential to bring an unusual level of public interest to bear on the industry itself.

What was really most significant about the moment has only begun to emerge in retrospect. When considered in the context of trends that can only clearly be traced to the early and mid-1980s, it appears increasingly possible that this industry—one of the largest, oldest, least innovative, and most important in industrial societies—has begun what could ultimately be a process of significant long-term restructuring. It has begun to demonstrate the first tentative moves toward potentially major innovations in materials and technology since the late nineteenth century.

The most visible changes in the past decade have been organized around wastepaper. Other, newer initiatives have focused on chemical process and energy use, while still others have been developing around forest management. As yet, however, many of these changes are occurring at the margins of the core industry. Despite occasional proclamations that the revolution has already occurred, the industry in the mid-1990s is in fact capitalized, organized, mechanized, and supplied in much the same ways that it has been throughout the latter half of the twentieth century. It supplies products that are largely indistinguishable from those it has supplied for many decades, to markets that continue to grow by leaps and bounds.

The truly significant difference—the one that suggests the possibility that a more deep-rooted process of transformation may have been triggered—has been in the way the industry is perceived by the society at large and in the degree to which it has been forced to respond to this changing perception. Most importantly, and for the first time in the industry's history, its internal decision making and its future are being substantially driven by a network of diverse and often conflicting environmental forces, ranging from federal pollution control mandates and international environmental accords to consumer education campaigns and grassroots mobilization around particular facilities, companies, forests, and products. Profit and competitiveness remain the industry's goals, but the major parameters are now environmental. Increasingly of interest with respect to the long-term trajectory of the industry are forces emerging at the intersection of environmental, social, and economic justice concerns and beginning to examine and test their potential at local and regional levels. To be able to evaluate the dynamics of these trends and to consider the issues associated with where such trends may lead, it is necessary to begin with an overview of the basic structure and characteristics of the modern industry and its markets.

Historical Development

As one group of analysts commented in recent years, "It is difficult to imagine any product with a more ubiquitous role in the civilization of developed, industrial economies than paper."[3] The extensive integration of paper products with both culture and economy reflects an industry whose historical roots are far-reaching indeed.

A vastly condensed outline of the world history of papermaking can be viewed in terms of roughly four historical stages. The first, which began in China around the first century A.D., saw early papermaking techniques based on vegetable fibers and rags as raw materials gradually progress westward from Asia to Europe. By the fourteenth century paper mills existed in Spain, Italy, France, and Germany. The second phase was inaugurated by the invention of mechanical printing in the mid-fifteenth century and continued until the end of the eighteenth century. It was a period characterized less by changes in papermaking technology than by the increasing rise in demand for paper that accompanied the developments in printing technology. This rising demand led to the search for new materials to supplement the cotton and linen rags providing the bulk of raw material for production and for associated technologies for processing new materials. The third phase, from the late eighteenth to the mid-nineteenth century, was the high era of technological invention for the paper industry—in fact the last era of truly significant invention and innovation until very recent times. The modern industry, the fourth phase, was founded with the commercialization and diffusion of these new production processes in the last half of the nineteenth century. This book is largely focused on the prospects for a fifth phase: a sustainable twenty-first century paper industry.

The eighteenth- and nineteenth-century technologies that laid the groundwork for the rise of the modern paper industry were focused on the use of wood as a new fiber source. They included the development of both mechanical grinding (groundwood) techniques, and various chemical methods for breaking down wood structure to create pulp. The first commercial applications of the new technologies began with groundwood pulping, which was underway in the United States by about 1840 and by the 1880s provided most of the paper used by newspaper publishers. The groundwood technologies were followed by the commercialization of the dominant chemical pulping processes including the soda, sulfite, and sulfate (or kraft) processes, which originated in Europe and spread to the United States between 1850 and the turn of the century. The element chlorine, discovered in 1774, came into use as a pulp bleaching reagent in the 1800s. Its use, however, was not perfected until the rise of kraft pulping as the dominant chemical process in the twentieth century, so various hypochlorite compounds were most commonly used for bleaching throughout the nineteenth century. Although incrementally improved over the next century, these basic processes have remained the foundation of modern pulp production.[4]

The importance of the new woodpulping technologies, which opened up a plentiful raw materials source, was matched by the invention of the continuous sheet paper machine, patented in Europe by the Fourdrinier brothers in 1807. Paper that had been made one sheet at a time by manually dipping a frame with a screened bottom into a vat of pulp slurry (a highly dilute suspension of pulp in water) could now be produced on a moving screen belt at speeds of several hundred feet per minute. The most common modern paper machines (which run at speeds of several thousand feet per minute) are still known as "Fourdrinier machines."

By the 1870s, when sharply increasing demands for paper combined with the availability of the new technologies and the new raw materials source, paper production had largely shifted to virgin wood as its primary fiber base, and the modern era of large-scale paper production was inaugurated. In the United States more than 60 percent of the fiber used in paper production came from virgin wood by the early 1900s, although the use of rags, wastepaper, and certain annual plant fibers remained significant throughout the first half of the twentieth century.[5] While the average virgin woodpulp content in domestic paper has risen to as high as 80 percent or more since then and has remained the foundation of modern paper production, there have been complications and occasionally dramatic fluctuations in fiber content connected with the use of both wastepaper and different sources of wood and other fiber, which will be discussed later. Not surprisingly, however, by the turn of the century the paper industry had become closely entwined with the timber industry.

Most of the U.S. paper companies that dominate both domestic and international paper production were founded between 1870 and the 1920s. They include (roughly in order of paper products sales in 1993) such venerable giants as International Paper (1898), the largest paper company in the world; Kimberly-Clark (1907), which in 1995 announced plants to purchase Scott Paper (1870); Georgia-Pacific (1927), which recently absorbed Great Northern Nekoosa (1874); Stone Container (1926); Mead (1879); Champion International (1929); Weyerhaeuser (1900); Union Camp (1874); Westvaco (1888); and Boise Cascade (1920). The top fifteen companies are rounded out by Sonocco, James River, and Jefferson Smurfit. With the exception of the James River Corporation, which rose during the 1970s by focusing on high-grade and specialty papers, it is only recently—indeed, since about 1990—that the top ranks have been broken by such upstart newcomers as Jefferson Smurfit, which among other enterprises has built a recycled newsprint and paperboard empire. Altogether, these fifteen companies accounted for about 40 percent of total U.S. paper industry sales in 1993.[6]

Industrial Organization

The modern pulp and paper industry is classified by the U.S. Department of Commerce as the "Paper and Allied Products Industry" under the two-digit Standard Industrial Classification (SIC) code 26. Seventeen distinct industries are described within this sector, although the organizational structure can be broadly divided into three stages, or tiers of activity: pulp production, paper and paperboard production, and finished products conversion. The various industries and their associated SIC codes are listed in table 1.1.

The first tier is comprised of *pulp mills* (SIC 2611), which engage primarily in the conversion of wood chips or logs into wood pulp. They

Table 1.1
Primary and secondary wood-processing industries, 1992

SIC code	Industry group and industry	Value of shipments (million $)
24--	Lumber and wood products	81,798
241-	Logging	13,844
* 2411	Logging	13,844
* 242-	Sawmills and planing mills	23,210
* 2421	Sawmills and planing mills, general	21,045
* 2426	Hardwood dimension and flooring mills	2,031
* 2429	Special products sawmills, n.e.c.	134
243-	Millwork, plywood, and structural members	24,865
+ 2431	Millwork	9,614
+ 2432	Wood kitchen cabinets	4,995
* 2435	Hardwood veneer and plywood	2,257
* 2436	Softwood veneer and plywood	5,516
+ 2439	Structural wood members, n.e.c.	2,482
244-	Wood containers	2,922
* 2441	Nailed wood boxes	445
* 2448	Wood pallets and skids	2,134
* 2449	Wood containers, n.e.c.	343
245-	Wood buildings and mobile homes	6,634
+ 2451	Mobile homes	4,532
+ 2452	Prefabricated wood buildings	2,101
249-	Miscellaneous wood products	10,324
* 2491	Wood preserving	2,677
* 2493	Reconstituted wood products	3,944
* 2499	Wood products, n.e.c.	3,703
25--	Furniture and fixtures	43,688
251-	Household furniture	20,707
+ 2511	Wood household furniture	8,762
+ 2512	Upholstered household furniture	6,221
2514	Metal household furniture	2,116
2515	Mattresses and bedsprings	2,843
+ 2517	Wood TV and radio cabinets	315
2519	Household furniture, n.e.c.	450
252-	Office furniture	8,002
+ 2521	Wood office furniture	1,961
2522	Office furniture, except wood	6,041

Table 1.1 (continued)

SIC code	Industry group and industry	Value of shipments (million $)
253-	Public buildings & related furniture	4,140
2531	Public buildings & related furniture	4,140
254	Partitions and fixtures	6,569
+ 2541	Wood partitions and fixtures	3,125
2542	Partitions & fixtures, except wood	3,444
259-	Miscellaneous furniture and fixtures	4,270
2591	Drapery hardware, blinds and shades	1,916
2599	Furniture and fixtures, n.e.c.	2,355
26--	Paper and allied products	132,954
261-	Pulp mills	5,457
* 2611	Pulp mills	5,457
262-	Paper mills	32,817
* 2621	Paper mills	32,817
263-	Paperboard mills	16,126
* 2631	Paperboard mills	16,126
265-	Paperboard containers and boxes	32,577
+ 2652	Setup paperboard boxes	436
+ 2653	Corrugated and solid fiber boxes	19,681
+ 2655	Fiber cans, drums, and similar products	1,922
+ 2656	Sanitary food containers	2,491
+ 2657	Folding paperboard boxes	8,047
267-	Miscellaneous converted paper products	45,977
+ 2671	Paper coated and laminated, packaging	3,719
+ 2672	Paper coated and laminated, n.e.c.	7,646
+ 2673	Bags: plastics, laminated, and coated	5,706
+ 2674	Bags: uncoated paper and multiwall	2,838
+ 2675	Die-cut paper and board	2,005
+ 2676	Sanitary paper products	15,468
+ 2677	Envelopes	2,838
+ 2678	Stationary products	1,401
+ 2679	Converted paper products, n.e.c.	4,357

* = primary processing; + = secondary processing

Source: U.S. Dept. of Commerce, *Census of Manufactures* (1992).

are broadly characterized as either chemical or mechanical mills depending on the primary process employed, although they may engage in both types of production or use hybrid technologies. Pulp that is dried in rough sheets and sold to off-site paper mills on the open market is known as *market pulp,* and represented about 15 percent of total virgin pulp produced in 1993.[7]

Pulp mills may also produce pulp from plant materials other than wood, collectively referred to as *nonwoods.* Nonwood fiber sources include agricultural residues from the harvest of food and seed crops such as wheat straw, corn stalks, rye grass straw, and others; crops grown primarily for fiber (industrial fiber crops) such as kenaf and hemp; wild reeds and grasses such as bamboo and esparto grass; and industrial or postconsumer wastes such as cotton ginning by-products, textile wastes, and by-products of sugarcane processing (bagasse). Within industrialized nations, nonwood fiber pulping has remained almost totally undeveloped by the modern paper industry, although within developing nations nonwoods represent roughly a third of all fiber sources used in making paper.[8] The potential for nonwood fiber use as a substitute for wood in the paper and other forest products industries has received increasing environmental interest in recent years and is discussed at more length in subsequent chapters.

Stand-alone mills primarily engaged in producing pulp from wastepaper (of which the first in North America began operating in late 1991) are also classified within SIC 2611 for government accounting purposes, although this designation masks substantial differences in physical plant and industrial process between virgin and recycled pulp production.[9] More frequently, wastepaper pulping capacity is added to an existing virgin mill. In this context, the wastepaper pulping facility will commonly be referred to as a *deinking plant* when it produces paper-grade (white) pulps and as a *recycling line* or *wastepaper plant* in the case of paperboard (brown) grades of recycled pulp. The comparatively minimal processing involved in the latter case makes it more analogous to conventional paper mill stages of stock preparation than to the intensive processing characteristic of woodpulping.

The next tier of paper production is made up of *paper and paperboard mills* (SIC 2621, 2631). *Paper* is both a broad term that generically refers to a variety of pulp-based products (including paperboard) and a specific term that differentiates between the two major classes of bulk production—paper and paperboard—on the basis of thickness. Molded pulp mills, distinguished by producing formed products such as egg cartons, are a third category within this tier although current production is negligible. Building paper and board mills, which produce specialized construction-grade products such as cellulose insulation and roofing papers, are another general class of mill that may engage in one or both aspects of the conversion of pulp to paper and paper plus other materials into building products (crossing over into the third tier of paper converters). Their total output, however, is vastly overshadowed by total paper and paperboard production, and is declining.

Paper and paperboard mills are primarily characterized by the presence of one or more huge paper machines (up to 20–30 feet wide and several hundred feet long) that consume a *furnish* of pulp, water, and various additives at one end and produce bulk rolls or sheets of paper or paperboard stock at the other. When pulp and paper mills are located at the same site, and production processes can run more or less continuously from raw materials (whether wood, wastepaper, or other fibers) to bulk paper or paperboard products, they are called *integrated mills*. They may be variously classified by SIC code as pulp, paper, or paperboard mills depending on the dominant end product (some integrated mills also produce market pulp). A by-product of the current industrial classification scheme is that systematic, comprehensive analysis of pulp mills from government economic and environmental data series is impossible. Manufacturing complexes classified primarily as pulp mills (SIC 2611) represent only a fraction of all pulp mills in operation, the majority of which are integrated with (and thus classified as) paper and paperboard mills.[10]

Both large-scale integrated mills and large-scale market pulp mills are the cornerstones of modern paper production and are invariably located close to the raw materials source. Until recently this has meant that large mills are located in or near rural forestlands, but the distribution of large-format pulp and integrated paper production has shown at least some potential for shift along new dimensions including rural/urban shifts around wastepaper. What, exactly, qualifies as "large scale" has been subject to continuous redefinition as paper production has become unremittingly more concentrated.

The larger integrated mills of the 1960s and 1970s, for example, which had one or two paper machines and produced perhaps 200,000–250,000 tons per year (tpy) of paper, have been replaced by a standard minimum capacity for new mills *(greenfield mills)* or major upgrades of existing mills, in the range of 300,000–600,000 tpy—approximately 800–1,500 tons per day (tpd).[11] Few integrated virgin paper or paperboard mills exist in the United States today that operate at levels much below 500–600 tpd. Most facilities with kraft pulp mills operate at levels of 1,000–2,000 tpd, as do many mechanical pulp mills. Recent years have seen the rise of a few extremely large-scale mills that operate at vastly greater production levels. The "million ton club" colloquially refers to the elite but growing supertanker class of mills with capacities at or above one million tpy (approaching 3,000 tpd).

As of the early 1990s, the largest pulp mill in the world was the Aracruz Cellulose mill, located in the Brazilian state of Espirito Santo. With a capacity of 1,025,000 tpy, it was the first member of the million ton club.[12] Major capital investment programs undertaken at several U.S. mills since then have (or soon will) put them in the same league. These huge enterprises have invariably been based around kraft woodpulp mills; however, in 1991 in a rather astonishing development, a 100 percent recycled pulp and paper mill was proposed for an urban area of Northern California with a target capacity of more than 750,000 tpy. (This project is the subject of chapter 5).

Unintegrated paper and paperboard mills are also an important component of the organizational structure of the industry, comprising roughly two-thirds of all paper and paperboard mills by number of facilities, although they are characterized by lower production levels than integrated mills. They typically produce stationary, publishing, or tissue papers, miscellaneous grades of packaging board or paper, and specialty grades such as filtration or cigarette papers from various blends of market pulps. With the latest upward inclination in wastepaper utilization, which began around 1986, the term *semi-integrated paper mill* has come into use to describe standalone paper mills that have "backwards integrated" to meet part of their pulp requirements through on-site production of recycled pulp. Unintegrated and semi-integrated paper(board) mills are relatively flexible in terms of location and are usually situated in urban or suburban areas in proximity to their markets. The scale of production typically varies from small to moderate—from about 200–600 tpd, although both smaller and larger unintegrated or semi-integrated mills exist.[13]

Since the early 1990s, the *minimill* concept has swept into vogue in the United States, led initially by tissue paper manufacturers. The term is used to describe new small-scale urban mills (typically about 100–300 tpd, although some as small as 20 tpd) that produce 100 percent recycled products. Deinked-grade minimills are usually organized around the use of office wastepaper (OWP) to produce recycled market pulp or are integrated with a paper machine for the production of tissue papers (which lend themselves to 100 percent recycled content). Paperboard-grade minimills consume old corrugated containers (OCC) and are usually integrated with a machine for the production of linerboard and/or corrugating medium.[14] These mills tend to be located in urban industrial parks where they capitalize on existing urban transportation, power, steam, water supply, and wastewater treatment infrastructure—unlike wood-based mills, which often own their own railroads, energy plants, and wastewater treatment plants and which draw their water supply directly from rivers and lakes. As recycling mills, they have far lower levels of gross energy, water, and chemical use. They are based on well-established technologies and strongly characterized by simplicity of design and process with a single fiberline devoted to producing a single product. Of equal interest, they have also tended to be independent mills without ties to major paper companies. By the end of 1994, more than forty such projects with startup dates in the mid- and late 1990s had been announced or undertaken.

It is too early in the development of this trend to know whether urban minimills may come to represent a significant form of restructuring and to capture a significant share of total production, or whether they are a niche phenomenon arising in the backwash of a short-term wastepaper supply/utilization imbalance. Unquestionably, they challenge the most fundamental premise of modern paper production, that bigger is better, and it is therefore not surprising that they have been developed almost exclusively by players from outside the established industry. Although these new mills represent an interesting and significant development, it is important to bear in mind that pulp and paper production remains overwhelmingly concentrated in large rural or semirural wood-based mills, geographically isolated from most paper consumers.

Virgin woodpulp mills are the nucleus of the modern paper industry. For the past century and a half these mills have followed the primary raw material base—the forest—around the country in the hunter-gatherer mode that has long characterized the forest products industries as a whole. Increasingly, the industry's natural resource perspective has taken on agriculturalist overtones as the era of tree plantations has risen. The axes of change in the industry with respect to the format of individual mills and their geographic distribution thus include not only the new urban/rural and large/small dimensions influenced by wastepaper (and perhaps increasingly complicated by nonwood plant fibers), but also the more traditional geographic shifts that follow the forests, as well as the increasingly significant issues of international trade in forest products.

What we have at present in the United States is an industry whose primary manufacturing format is large and rural, with a more modest counterpart (by total production, although much larger by number of establishments) that is small and urban; both formats are growing in capacity. There are currently about six hundred paper and paperboard mills and about three hundred and fifty pulp mills in the United States. More than 80 percent of pulp mills are integrated with paper or paperboard mills, whereas more than half of all paper and paperboard mills are unintegrated.[15]

In Canada, by way of contrast, pulp mills slightly outnumber paper and paperboard mills. This fact reflects a number of issues that both differentiate and integrate the U.S. and Canadian paper industries. They include the Canadian industry's concentration in lower value-added commodity product areas such as integrated virgin newsprint production and its high levels of pulp exports (about 40 percent of total pulp production), of which about half are to the United States.[16]

The topmost tier of the paper industry, of only limited importance to this discussion, consists of roughly six thousand converting establishments, which convert bulk paper and paperboard stock into finished products. They are classified primarily within SIC codes 267 (paper converters such as envelope and paper bag manufacturers) and 265 (paperboard container manufacturers).

Within the paper and allied products sector as a whole there is a very high level of vertical integration: most paper and paperboard commodity sales originate from companies that also produce woodpulp and own or control timber resources. It is probably the single most recognized characteristic of the industry.

The Forest Products Industries

Missing from the picture of the pulp and paper industry painted by SIC codes is, of course, the forest. To get an accurate view of the industry in relation to its primary raw materials base and in terms of the structure of the leading paper producers, it is necessary to view it within the larger context of U.S. forest products or wood-based industries.

One way to view the wood-based industries is from a materials-flow perspective that begins at the forest. The MacArthur Foundation, for example, in an effort to provoke more systemic thinking about the potential for sustainable forest management recently produced a matrix outlining the major issues associated with each of a chain of significant phases that link the forest to the end-consumer of wood-based products. The phases are: resource management, extraction, primary processing, secondary processing, wholesale, retail, and end use.[17] The paper and allied products manufacturing sector described above includes primary (pulp and paper mills) and secondary processing (paper conversion) establishments.

In addition to pulp and paper mills, the other major route by which wood flows from the forest into the economy is via sawmills and wood panel mills. Within the system of SIC codes, the sector in which these industries are housed is known as the "Lumber and Wood Products" group (SIC 24). Seventeen industries are described within this sector, but each falls generally within one of three areas: logging (extraction), sawmills and wood panel mills (primary processing), and miscellaneous wood products manufacturing (secondary processing), such as wood cabinet manufacturers. In addition, various wood-based sectors of the

"Furniture and Fixtures" group (SIC 25) round out the secondary wood processing industries that build on the output from sawmills and panel mills (see table 1.1).

Logging, which of course underlies all the forest products industries, is nevertheless classified within the lumber and wood products sector (under SIC 2411). (Timber production and the ownership and management of timberland are discussed in the following chapter.) *Sawmills and planing mills* (SIC 242) produce lumber and other solid wood products such as wood flooring and shingles and are, together with pulp mills, the most important of the wood-based industries by volume of timber consumed.

The *wood panel* industries are dominated by plywood and veneer mills (SIC 2435, 2436). Increasingly important, however, are manufacturers of particleboard, oriented strandboard (OSB), medium density fiberboard (MDF), and related panel products (SIC 249).[18] The wood panel industries, along with glue-laminated (glulam) lumber manufacturing, are also generally known as *engineered wood products* industries since they create composite materials by laminating, molding, or otherwise reconstituting wood fragments with various binders and other additives. Their relationship to wood supply is similar to pulp mills' in that although they may consume logs, many are partly or wholly dependent on wood chips and particles produced at sawmills. As with pulp mills, nonwoods and wastepaper constitute increasingly feasible alternative fiber sources for the manufacture of both structural and nonstructural "wood" panels. The similarity between these two parallel sectors of the forest products industries (i.e., between engineered wood panel producers and pulp mills) is important. As discussed in the next chapter, a future-oriented perspective views them both as *lignocellulose-based materials industries* that may or may not have wood as their primary fiber source.

The secondary wood processing industries, which build upon solid and composite wood commodities, encompass a diverse array of manufacturing activities, including the production of furniture and cabinets; prefabricated windows, doors, moldings, stairs, and trusses; and prefabricated wood buildings such as mobile homes. They bear the same relationship to sawmills and panel mills as do paper converters to pulp and paper mills: they do not place a primary demand on the forest, but rather consume intermediate industrial commodities, and they may be integrated with primary wood processors in terms of both corporate organization and plant format.

In 1992, the U.S. *Census of Manufactures* counted 13,000 logging establishments (up by nearly 10 percent since 1987), 7,068 sawmills (up 5 percent since 1987), and roughly 1,500 mills producing panels and similar engineered wood products. Secondary wood processing establishments numbered more than 20,000 (including selected industries within the two major sectors described by SIC codes 24 and 25).[19]

One difficulty with the materials flow model identified above (undoubtedly recognized by its creators) is that it dramatically oversimplifies the world of end consumers. For example, is the newspaper publisher the end consumer, or is it the newspaper reader who purchases an information product, but last touches the material before it is disposed of or recycled? There are obvious and important differences between end consumers of newspapers and houses and intermediate industrial consumers of newsprint and lumber. For basic commodities such as lumber, panels, and paper one or more major industries typically stand between the final consumer of the material, and the primary and/or secondary wood-based industries.

Various forms of horizontal and vertical integration may also exist between the commodity materials industries and their various industrial customers. Companies engaged in producing wood products are often horizontally integrated with the manufacture of other building products, and it is not unknown to find major wood-based corporations with divisions and subsidiaries engaged in construction and development. Likewise, it is not unknown to find publishers with holdings in pulp and paper mills. Thus, in addition to the obvious class of secondary wood-processing industries, a range of other important industrial sectors are closely connected to and sometimes integrated with the dominant *primary timber processing industries:* pulp and paper mills, sawmills, and wood panel mills.

Most woodpulp mills are owned by large, publicly traded corporations. Sawmills and panel mills are predominantly owned by either public or private corporations. Logging establishments, however, are typically small, single-unit, and non-corporate in organization, and usually operate as independent contractors to primary processors. In the southern United States, for example, where pulp production has become strongly concentrated, a somewhat unique form of this industrial organization developed around the complicated timberland ownership structure characteristic of the region. The contractors who log trees destined for pulp production (pulpwood) are known as pulpcutters and have traditionally operated in a profoundly oppressive economic relationship with paper companies. Independent dealers (or independently operated divisions of paper companies) operate woodyards capitalized by paper companies and work as middlemen between the landowners, the pulpcutters, and the paper company. They typically negotiate for stumpage rights and then hire pulpcutters, who are paid a piece rate for each cord of wood, to cut and haul the wood to either the woodyard or the company mill.[20] In other cases, where large tracts of public or industry-owned timberland is involved, logging contractors may work directly with the paper or lumber company to log specific areas to which the companies have secured cutting rights.

In their capacity as contractors, and as the most visible agents of the connection between the forest and the wood-based manufacturing industries, log dealers, haulers, and loggers provide something of a buffer between the industries placing a primary demand on timberlands and the actual logging of trees. One by-product of this form of industrial organization is that it is extremely difficult to follow the trail from raw materials to manufacturer. Although it is difficult to follow the wood from the forest to the sawmill, it is nearly impossible to follow the trails to the pulp mill. Despite the fact that lumber and pulp production account for most of the demand on domestic forests, remarkably little information connects specific logging operations with specific companies. The question "how many trees" is thus quite different from "which trees." Paul Ellefson and Robert Stone, who analyzed the structure of the U.S. wood-based industries in their book *U.S. Wood-Based Industry: Industrial Organization and Performance* (1984), ran into similar problems in attempting to evaluate the corporate ownership of forestland: "Determining industrywide control of land and the products produced thereon is one thing—defining specific corporate holdings is another. . . . The continuing void of information on ownership and control of forest land is surprisingly large."[21]

The familiar "owls versus loggers" representation of western forest conflicts is one indicator of the remarkable degree to which the large wood-based corporations have remained insulated from a public perception of their responsibility for destructive logging and forestry practices. Loggers, who undeniably make for a more colorful story than corporate executives, are nevertheless only hired hands for more powerful economic interests. Yet, the extreme complexity of documenting the connection between particular corporations and products, and particular forest areas and types of logging practice, is one of the barriers to a more accurate understanding of the industry and its major players.

Lately, several environmental organizations have honed in on this murky chain of relationships. Although efforts focused on particular forests and forest products companies or particular types of products are longstanding, recent campaigns have gone further by following the materials chain from logging operations to high-visibility corporate end consumers, in some cases using data obtained under state public records acts. Rainforest Action Network and Greenpeace, for example, have targeted customers of MacMillan Bloedel (a huge Canadian forest products company) such as the *New York Times,* GTE, and Pacific Bell, highlighting their products' (newspapers/phone directories) direct material line of descent from MacMillan Bloedel's clearcut logging of pristine rainforests on Vancouver Island, British Columbia.[22]

Although the production of paper and lumber or other wood products involves distinct industries, they are often housed within the same company. Of the leading paper companies identified above, most are also among the largest lumber producers in the country. These large, *integrated wood products corporations* include International Paper, Georgia-Pacific, Boise Cascade, Mead, Champion International, Weyerhaeuser, Union Camp, Westvaco, and Temple-Inland. Georgia-Pacific alone had more than $4 billion in lumber and wood products sales in 1990, or almost 6 percent of total sales by this sector, and more than 5 percent of total paper products sales. Other major companies prominent in both paper and lumber sales include Federal Paperboard, Willamette Industries, Manville, Potlatch, Simpson, ITT, and Chesapeake: each had between $700 million and $2 billion in paper sales, and between $100 and $600 million in lumber and wood products sales in 1990.[23]

The vertical integration of the paper industry is much remarked, but the level of horizontal integration with other wood-based industries is less frequently mentioned. From an alternative materials use perspective—one interested in the substitution of wood use by other fiber sources including both wastepaper and nonwoods—it may be even more important because it emphasizes the degree to which a complex and far-reaching industrial infrastructure has been defined and elaborated by its relationship to a wood resource base.

Although most of the major players in the wood processing industries have lumber, paper, or both as their primary business, they also, as suggested above, engage in a wide variety of other businesses. Notable are companies with subsidiaries or independently operated divisions involved in land management, transportation, chemicals, utilities, and development and construction—generally businesses that either supply, distribute, or consume services and products used and produced by integrated wood-products corporations. Georgia-Pacific, for example, owns seven railroad companies, a steamship company, two power companies, a chemicals company, and dozens of paper and paperboard converting companies, among its seventy or so subsidiaries. Weyerhaeuser, like several of the leading U.S. wood products manufacturers, has major subsidiaries operating production facilities in Canada, and extensive sales and distribution operations in dozens of foreign countries. It owns more than 5.6 million acres of timberland in the United States, a savings and loan association, several insurance companies, and other financial services businesses. Many of the major U.S.-based companies are involved in joint venture production activities in developing countries.

Total sales by each of the top fifteen U.S. paper companies ranged between $1.6 and $13.6 billion in 1994. Paper and allied products sales accounted for 78 percent of all sales by these top producers, ranging from 38 to 100 percent by individual company.[24]

Thus, not only is domestic paper production substantially concentrated within fifteen to twenty very large transnational corporations, but most of these corporations are themselves intricately embedded within the structure of the forest products industries as a whole. In terms not only of access to wood, but also of the subsidiary activities of these corporations in areas such as transportation, chemicals, and utilities, which often serve or are otherwise integrated with primary production facilities, production capacity has become strongly associated with specific locations, particular production formats, and a variety of strong inter-sectoral dependencies.

It is difficult to overstate the significance of how the forest products infrastructure and its linkages to external sectors supports and defines the corporate and organizational structure of the paper industry. Equally important is the degree to which it influences the paper industry's capacity to innovate, its interest in innovation, and the paths of change it may be likely to pursue. The specific implications for the use of alternative fiber sources are unmistakable: taking the wood out of paper may well be technologically feasible, but taking the paper industry out of the forest is something else entirely.

Reaffirming and reinforcing the economic and political strength of this relationship, the American Paper Institute merged in the early 1990s with the American Forest Products Association, to become the American Forests and Paper Association (AFPA), based in Washington, D.C. The symbolic importance of this joining of forces is difficult to miss, especially at a time when the use of wood by both sectors was beginning to be actively challenged by a new expansion of the traditional forest conservation agenda, as will be discussed later. Paralleling the forces that arise from the complex integration of the paper industry with the other forest products industries, however, is an equally important set of forces that arise from the sheer capital scale of modern paper production.

Capital, Scale, and Innovation

As the globalization of the economy rises, the study of technological innovation has become more prominent than ever in the study of long-term economic growth strategies. Couched in the language of technological and industrial competitiveness, and often within the context of national technology policies or industrial policy debates, modern innovation studies have helped define the development of new high-tech capabilities in areas such as biotechnology, new materials, as well as microelectronics and information technology as core elements of national economic stimulus.

Mature industries faltering in the face of increased foreign competition or fleeing in search of cheap labor and weak regulation have also been subject to the scrutiny of innovation analysts and given a place in the discussions over technology and (especially) industrial policies. At the same time, developments in various decentralized and modular formats for flexible production, coordinated across continents by strategic brokers at the center of enterprise webs (facilitated by new information technologies), have been heralded in such widely read books as Robert Reich's *The Work of Nations* (1992), as well as in the drier academic and policy literature. Innovation—particularly technological innovation—for purposes of competitiveness clearly continues to be a central focus of economic policy, and the more innovation the better seems to be the guiding principle.

In the enthusiastic pursuit of advanced technologies, accelerated innovation in established technologies, and new formats for organizing production, the prospect for substantive innovation within the paper industry has tended to be a somewhat marginal issue. As a mature industry, it is by definition characterized by decades of successful competition, highly optimized production technologies and formats, and relatively stable products and markets. Because it is neither faltering nor fleeing, the internal and external processes by which innovation in the industry is induced, forestalled, or directed have received little attention.

Within the past decade or so, this has begun to change due to two factors. The first is that the policy analysis of technological and organizational innovation, industrial competitiveness, and economic growth has been rhetorically joined in a few segments of the prevailing literature with an expressed concern for the environment. Particularly since the publication of the Brundtland report (as the 1987 report by the World Commission on Environment and Development is commonly known), it has become at least somewhat inappropriate for political figures to make major public policy pronouncements on innovation and competitiveness without a footnote or two on environment and sustainability. This, in turn, mildly resuscitated interest in revitalizing processes of innovation in industries like the paper industry, which, although competitive and stable, are also associated with profoundly negative environmental impacts.

The second, more important reason for the growing interest in innovation in the paper industry is less rhetorical and more conventional, although also strongly related to environmental issues. It arises at the intersection of the extreme structural rigidity of the U.S. paper industry and increasing global trade, at a time when environmental concerns about the industry have come to a boil in both North America and Europe. If the domestic industry is not yet faltering or fleeing, declining competitiveness has become a possibility. The increasing costs associated with meeting rising standards for environmental performance in the United States have been accompanied by the expectation of increasing inroads on the U.S. and world markets. Of new importance are producers in developing countries such as Brazil, Chile, Indonesia, and others who are operating new mills with low labor costs, and substantially more limited environmental regulatory oversight. Not surprisingly, technological innovation is viewed as the talisman that will ward off the challenge they pose.

The two perspectives—(1) technological innovation within the largely rhetorical context of sustainability, and (2) innovation for purposes of economic competitiveness—have lately achieved an odd merger with respect to the paper industry because the economic issue is significantly being driven by a changing environmental framework. In a sense, the rhetoric of the former (sustainability) has been adopted for purposes of advancing the latter (competitiveness). This of course assumes that innovation oriented toward long-range goals of social, ecological, and economic sustainability is inevitably consistent with innovation oriented toward short- and mid-term goals of economic competitiveness by the established industry when competitive strength has an environmental dimension. This assumption must be questioned.

Before returning to the paper industry's capacity for innovation, it is worth further outlining some of the dominant themes and approaches characteristic of the ways in which the subject of the environment is generally linked to the subject of technological innovation in the prevailing technology and competitiveness debates.

The first approach might be considered the *blind faith* approach. This view assumes that the technological innovation we are chasing will probably produce cleaner and socially friendlier technologies, for good sound economic reasons. The hidden foot of some form of on-going environmental regulation will be a continuing influence. The natural pursuit of greater production efficiency—making more with less—will be the other major influence. This widely shared point of view is typified by Christopher Freeman and Geoffrey Oldham in their introduction to *Technology and the Future of Europe: Global Competition and the Environment in the 90s* (1991): "Technologies which improve the quality of life, which protect the environment and which are energy- or material-saving, may well become the cutting edge of world technological competition." It is similar to the even more generic optimism of 1980s- and 1990s-style U.S. environmental posturing typified by the remarks of former U.S. EPA Administrator William Reilly: "In the newly emerging marketplace, the green of environmental protection is beginning to form a ready alliance with the green of profits."[25]

If we are wrong or uncertain about the social and environmental forecast for new technologies, there is a corresponding level of faith in the institutions we have developed in which to debate problems that arise: "Since there is no way that the market can assess or anticipate many of the future social costs and benefits of technical change some form of political debate and decisionmaking is unavoidable . . . The best that can be hoped for is to disclose some of the possible and probable consequences of technical change and to stimulate a widespread public debate among those who may be affected for good or for ill."[26]

A second, related approach is the equally generic *surplus capital approach*. In this view technological innovation is good for people and the environment in large part because it is good for the economy. Thus, as articulated in the last decade by a group of mainstream environmental leaders in the report *America's Economic Future: Environmentalists Broaden the Industrial Policy Debate* (1984), "Past environmental gains will be maintained and new ones made more easily in a healthy economy than in a stagnant one where dissatisfaction and unemployment reinforce pressures to cut back on environmental standards." As argued by the environmental leaders, this approach is somewhat more reliant upon an environmental regulatory apparatus than on the natural evolutionary benefits of greater production efficiency. They observed, "[Although] the evolution of industrial technology is in the direction of lower-polluting processes, the improvement is uneven and slow. Hence, there are likely to be significant efforts to commercialize environmentally benign production processes only if . . . prodded by the prospect of continued stringent regulation."[27] The Brundtland Commission shared the belief that environmental agencies must be given more power to "cope with the effects of unsustainable development" and that technological innovation was key, both for reducing poverty (and thus social instability that works against the environment) and because of the natural tendency of industries to innovate in the direction of producing more with less.

A third, overlapping approach is the ambiguous *environmental technology approach*. It is ambiguous because, arguably, all technologies have some environmental effect and are thus "environmental." The term is often conflated with other terms such as *sustainable technology* and *clean technology*. The Clinton administration's technology policy, for example, partially summarized in the National Science and Technology Council's report *Technology for a Sustainable Future* (1994), defines an environmental technology as one that "advances sustainable development," although it admits that sustainable development is itself hard to define. The categories into which the report divides environmental technologies, however, point us toward the operational meaning of the term. The categories are avoidance, monitoring and assessment, control, and remediation and restoration. They are, in short, environmental cleanup technologies. What's more, they are a big growth industry with enormous export potential.[28]

Some analysts have approached the margins of the major debates on technological innovation and the environment from the perspective of a more proactive and strategic environmental regulatory platform. This approach, particularly as articulated by Nicholas Ashford and associates, relies on a sophisticated understanding of sectoral dynamics to support focused regulatory interventions based on an attempt to stimulate particular types of technological innovation and diffusion. From this point of view, environmental regulation, rather than being a sort of passive, fail-safe backdrop to economic growth and technological change, becomes an active tool for stimulating fundamental change in industries.

In their article "Regulation and Technological Innovation in the Chemical Industry" (1983), Nicholas Ashford and George Heaton, for example, argue, "It should be acknowledged that the long term purpose of all the regulatory systems applicable to the chemical industry is to change, in a fundamental way, the nature of the technology the industry employs." It follows, as they point out, that the process will be disruptive: "Individual regulations will inevitably displace the market position of particular entrenched technologies" and, by extension, the structures that have built up around them.[29]

Their approach represents one vanguard of excursions from the mainstream environmental regulatory camp into the uncertain terrain of integrated economic and environmental policy. One of the things that most distinguishes it from the other approaches outlined above is the requirement for a substantially more sophisticated analysis of industrial sector structure as the context in which technological innovation is approached from an environmental perspective. Unlike the current focus of technology policy on new core technologies as the locomotives of economic growth, with presumed spillover advantages for environmental concerns, this approach is fundamentally concerned with the structure and technology of mature and stable sectors that nevertheless pose continuing and significant environmental problems.

This focus on the contextual factors that surround and flow from technological innovation has become increasingly important in the larger debate on national economic competitiveness. Donald Hicks, for example, in the book *Is New Technology Enough: Making and Remaking U.S. Basic Industries* (1988), argues that the model of technology-led industrial change and economic growth must be more qualified, and that the importance of technological innovation in and of itself is receding.[30]

From whatever perspective one approaches the subject of technological innovation, industrial competitiveness, and the environment, it is clear that the capacity for change is central to the discussion. However, the traditional economic logic of industrial growth patterns and the behavior of mature industries tends to work in the opposite direction: toward stasis rather than change. The industry's internal capacity and tendency toward technological and organizational innovation must thus, as Ashford suggests, be considered before the effects of external influences can be well understood and interventions appraised.

As Ashford and Heaton summarize, "When technological innovation first became the subject of serious study, the innovation process was typically depicted as a sequence of more-or-less discrete steps." They were broadly construed as basic research, applied research, invention, prototype development, and commercialization and diffusion.[31] More dynamic models came to characterize technical innovation within the context of observed industrial growth patterns. Early, middle, and late phases of a particular industrial sector's development were seen to demonstrate characteristic changes in the technology and organization of production. In the work of William Abernathy and James Utterback, the parameters that describe this trajectory include the nature of the core technologies, the competitive emphasis, materials-use characteristics, product diversity, equipment and mechanization characteristics, and industrial plant format.

According to this view, when a productive sector organized around a particular new technology (or a set of related technologies) comes into being, it is characterized by a fluid stage in which, for example, production processes are generally flexible and inefficient, the physical plants are small-scale, and the competitive emphasis is on product performance. In the middle phase the rate of change in the product declines, but the level of innovation in production processes rises. In the mature or rigid phase the level of change in both products and processes has become limited to incremental improvement, and the competitive emphasis is on cost reduction. The physical plant has become large-scale and dedicated to specific products; production processes are efficient, highly automated, and highly capital-intensive; and vertical integration with raw materials inputs is likely to be extensive.[32] Characteristically, the number of firms involved by the middle phase declines by the mature phase as the average size of firms grows to meet the increasing capital requirements of mass production.

The modern pulp and paper industry is essentially the archetype of an industry in the mature phase. Its closest analogues are the basic chemicals, primary metals, and oil-refining industries. These heavy manufacturing or smokestack industries are all are old, resource-based chemical processing industries, dominated by a limited number of large manufacturers. Their core products have long-established patterns of consumption and integration. Yet, as Ronald Slinn points out in a short article with the enticing title "The Paper Industry and Capital" (1992), the paper industry is, in fact, the single most capital-intensive industry among major U.S. manufacturing sectors.

Over the past decade new capital expenditures as a proportion of sales have averaged 6.9 percent for the paper industry; twice the average of all manufacturing industries. Among primary processing mills (pulp, paper, and paperboard mills), new capital expenditures averaged more than 10 percent of sales. Among pulp mills, new capital expenditures spiked above 20 percent of sales during the first half of the 1980s, running at 10–15 percent by the late 1980s and early 1990s. The comparable figures for basic chemicals manufacturing (e.g., chlor-alkali plants) and oil refineries fluctuated between about 2 and 5 percent in the same period. A corollary to this level of capital commitment to the physical plant is the highest capacity utilization rate among all U.S. industries: an average of 92 percent over the past twenty-five years, relative to a manufacturing average of 82 percent.[33]

The efficiencies of the capital investment are reflected in the productivity of employees. In the last two decades (1972–92) total annual production increased by 42 percent. (For purposes of calculating labor productivity in primary processing establishments, total production is the sum of paper, paperboard, and woodpulp production). In the same period, total employment in the paper and allied products sector held roughly constant at about 667,000 and total employment in pulp, paper and paperboard mills *declined* by 6.3 percent. Growth in labor productivity (tons produced per worker) in primary processing establishments averaged 2.3 percent per year during the period. In the last ten years (1982–92), growth in labor productivity averaged 3.2 percent per year. Stated another way: although the total throughput of material increased by about 25 million tons in the last twenty years, employment in primary processing establishments declined by about 9,000 jobs. Net growth in labor productivity in the period was 52 percent.[34]

These trends were accompanied by the increased concentration of production in large facilities referred to earlier. Although total material throughput increased by 51 percent in the last two decades, the number of pulp, paper and paperboard mills decreased by 21 percent, and the average output per establishment (tpy) increased by 90 percent. (Total material throughput is tons paper and paperboard plus net pulp exports if greater than zero.) Much of the increased scale and concentration of production is accounted for by large kraft pulp mills. The average capacity of the now roughly 120 kraft pulp mills in the United States increased by almost 32 percent in the past ten years to approximately 1,200 tpd.[35]

Based on domestic and Canadian projects announced within the past several years, ballpark capital costs for common types of new (or replacement) pulp production capacity may range from about $150,000 per tpd capacity up to nearly $1 million per tpd capacity.

Uncomplicated new wastepaper processing lines (e.g., for newsprint or paperboard grades), using established technologies, have new capital costs usually ranging from about $200,000 to $300,000 per tpd capacity, although lower figures have been reported.[36] If substantial renovation or expansion of subsidiary facilities such as wastewater treatment systems is required, the cost may be higher—up to $350,000 per tpd capacity or more. New wastepaper processing capacity is typically added to existing mills in increments of roughly 200–400 tpd, thus the total new capital costs of newsprint or paperboard wastepaper processing lines typically ranges from about $50 million up to about $150 million. Coupling such fiberlines with a paper machine can double the cost. The reported cost of high-grade deinking plants ranges from about $350,000 to $550,000 or more per tpd capacity, depending on the deinking and water treatment processes and requirements.

New or replacement woodpulp capacity can be fantastically expensive as indicated by Canadian projects announced in the early 1990s. For example, a greenfield chemical woodpulp mill recently built in the province of Alberta by Alberta-Pacific Forest Industries had an announced capital cost of about (U.S.) $1.3 billion, or about $900,000 per tpd capacity. Another mill complex proposed for Alberta by Grande Alberta Paper (the subject of intense environmental opposition) would include a sawmill, cogeneration plant, pulp mill, and three paper machines, to be built over a decade at a cost of (U.S.) $2 billion or more.[37]

The capital barrier to change in the primary manufacturing industry is thus twofold. Existing pulp and paper manufacturers typically have several hundred million to more than a billion dollars invested in each of one to a dozen or more large, efficient facilities, which must operate more or less continuously at near capacity and are often substantially integrated with other capital-intensive facilities. They cannot easily modify such facilities unless the existing capital investment has been fully depreciated and because the facilities must run essentially full-time to be profitable (a one- or two-week shutdown of a large paper machine merits announcement in the trade literature). However, companies may rebuild old mills to improve efficiency and increase capacity.

Prospective new manufactures using alternative process technologies and new materials (and, presumably, fielding added research and development costs) must nevertheless compete with existing products and prices. Chlorine-free paper, or nonwood fiber paper, for example, is expected to conform to the same performance standards as chlorine-bleached or wood-based paper and must generally compete on the same price basis. Thus new entrants and innovators, unlike those in a fluid early stage of technological development and industrial growth, do not have the luxury of competing in a new product niche with other young companies on the basis of quality, but rather are pitched against large integrated manufacturers of standardized commodities in a price competitive market. Although they may bring significant advantages in terms of reduced pollution control costs and reduced raw materials costs, they also bring uncertainty and risk, which are very poor companions to entry costs that usually begin at $50-$100 million and can easily climb into the hundreds of millions.

Thus far, all but a handful of alternatives to traditional production formats and technologies built or proposed in the United States and Canada in recent years have in fact relied on existing technologies that were basically waiting in the wings for a shove—conventional newsprint deinking being the most prominent example—and that with abundant and secure ONP supplies carried little uncertainty and risk. Although some recent wastepaper-based projects such as urban minimills have creatively capitalized at a smaller scale on various advantages of urban industrial formats (including proximity to wastepaper supplies and paper markets, and existing support services such as transportation infrastructure, electric substations and co-generation plants, and publicly owned wastewater treatment works), they have largely served to expand domestic capacity at the margins rather than to rework it fundamentally.

The level and role of directed industrial research and development spending by both private industry and government represents another of various possible measures of innovative capacity and inclination in a particular sector. Although increased R&D support may be the single most obvious tool for government technology or sectoral policies, or for certain forms of regulatory policies, it is dogged by many complexities that include how to focus funding, who to give it to (industry or universities, big or small companies, public/private research consortia, etc.), and how to integrate it with broader economic and environmental planning perspectives and with existing generic forms of government influence on the industry (such as pollution control, labor, energy, and natural resource tax laws and regulations).

Nevertheless, R&D investment by both companies and government has become a classical measure of innovative capacity. It should not be surprising in light of the preceding discussion that R&D spending by the paper industry is the lowest among all manufacturing industries (roughly tied with the food and related products sector). In the past two decades (1970–91) company R&D investment within the paper industry (as a proportion of sales) averaged about 0.8 percent or about a third the average of all manufacturing (2.8 percent). Although company R&D spending generally increased or held constant among other major manufacturing sectors in the 1980s (increasing on average from 2.1 to 3.5 percent of sales), it declined in the paper industry during the same period (from 0.9 to 0.7 percent). Government funding for industrial R&D directed specifically toward paper production has been negligible, although this does not reflect investment in R&D funding of related sectors (forestry, for example) or relevant areas of basic and applied science (e.g., cell biology and biotechnology). In the past two decades, total R&D funds (both government and company supplied) in manufacturing industries have climbed from 3.7 to 4.7 percent of sales; in the paper industry total R&D funds declined to 0.8 percent.[38]

The industry itself acknowledges that "technological leadership, once clearly owned by the U.S. industry, has . . . shifted towards Canada and the Scandinavian countries over the past 20–30 years." The AFPA observes that the industry's "high capital intensity and the resulting economic consequences of equipment replacement tend to limit experimentation, development and application of large, new core technologies."[39] Accordingly, in the face of technological superiority among longstanding competitors, rising challenges from new competitors, and increased public and regulatory scrutiny, the industry recently joined with the federal government in a collaborative program of research and development. Initiated in late 1994 with a first-year federal commitment estimated at $45 million and disbursed through the Department of Energy, "Agenda 2020"—as the partnership with DOE is known—at first glance represents a significant departure from the trends of the past decades. The program is further considered in later chapters.

Production and Consumption

Total sales by the U.S. paper and allied products sector in 1992 were $133 billion. The value of shipments was roughly comparable to that of the primary metals sector ($138 billion), about 90 percent of the oil refining sector ($150 billion), and about 45 percent of the chemical and allied products sector ($306 billion). The lumber and wood products sector had sales of $82 billion. The value of shipments from the paper sector represented 4.4 percent of the total for the domestic manufacturing industries.[40] The recent merger of the API and the NFPA reflected the consolidation of forest products sectors that together represent more than 7 percent of the manufacturing contribution to the gross domestic product (GDP).

Domestic production of paper and paperboard products amounted to nearly 85 million tons in 1992. By 1994, the most recent year for which data is available, production had increased nearly 6 percent to 89.6 million tons. Over the past twenty-five years annual growth in total paper(board) production has averaged 2.5 percent, but has averaged above 3 percent in the past decade (since the mid-1980s). Franklin Associates estimates that production will rise to 100 million tons per year by the year 2000—a fourfold increase since mid-century (see figure 1.1).[41]

Exports have represented an increasingly important market for domestic paper and paperboard producers. During the 1970s and 1980s exports

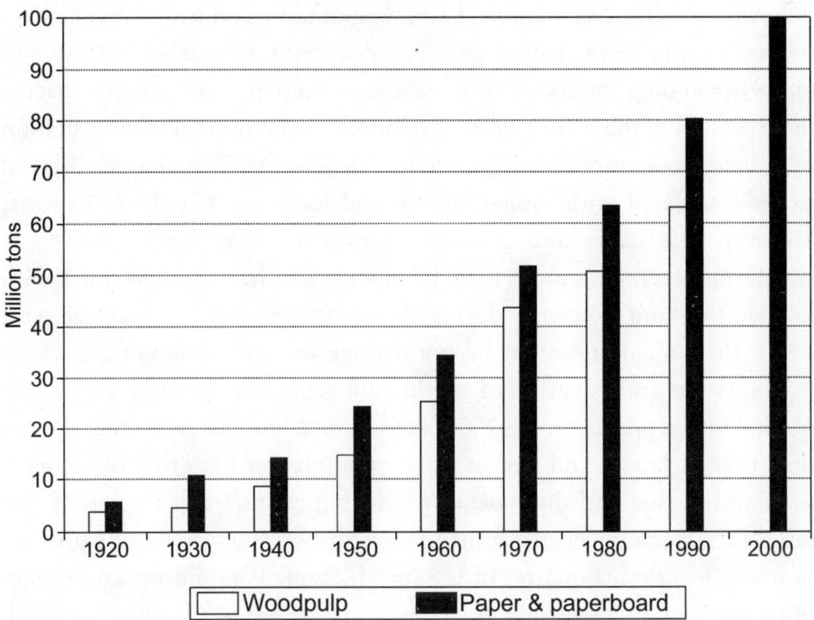

Figure 1.1
Woodpulp and paper production, 1920–2000
Sources: Tables 4.1 and 4.2; Franklin Associates, Ltd., *The Outlook for Paper Recovery to the Year 2000* (1993) (estimate for 2000).

accounted for 5–6 percent of total production (by tons). During the 1990s, however, exports have averaged more than 10 percent of production, reaching 11.5 percent (10.3 million tons) in 1994. Despite the growth in exports, the U.S. is still a net importer of paper and paperboard commodities, although the gap is closing. In 1994 imports of paper and paperboard, at 14.3 million tons, represented about 4 percent of total apparent paper(board) consumption in the United States, down from the 6–10 percent levels characteristic of the 1970s and 1980s. The single most significant reason for this decreased import dependency is likely the recent growth in wastepaper use as discussed in chapter 4. Canada is the most important trading partner: more than 17 percent of all exports were to Canada, and 83 percent of imports (dominated by newsprint imports) came from Canada.

Lower on the production chain, woodpulp production reached 64 million tons in 1994, having grown very little in the early 1990s due to factors including the focus on increased wastepaper utilization capacity, the recession of the early 1990s, developments surrounding the regulation of chlorine use in chemical woodpulping (discussed in chapter 3), and issues associated with timber supply and location. Woodpulp exports, however, like paper and paperboard exports, have been increasingly significant, averaging more than 10 percent of production in the 1990s, relative to about 6 percent during the 1970s and 1980s. Since the mid-1980s the United States has been a net exporter of woodpulp.

U.S. paper and paperboard production represented nearly 30 percent of total world production in 1991 and was one and one-half times greater than that of Japan, and almost three times greater than that of China—the number two and three paper producing countries in the world (see table 1.2). Canada was the fourth-largest producer, although its production was less than a quarter of U.S. production. Total European production was equal to 93 percent of U.S. production, led by Germany, Finland, and Sweden. The U.S.S.R. (primarily Russia) was the sixth largest paper producer in the world in 1991, although by 1994 Russian production had fallen by more than half due to transition problems since the collapse of the Soviet Union in 1991.

These eight countries, which together account for two-thirds of world paper production, represent very different dynamics in terms of how the paper industry is situated with respect to the national economy, natural resources, domestic consumption, and environmental regulation. Various details of these dynamics will be explored in more depth in connection with the domestic industry; however, the following provides a brief overview of the major international producers.

The United States and Canada both inherited vast forest resources, although both have made considerable inroads on their original forests. The United States has approached the limits of reliance on virgin or naturally regenerated forests and is moving into a new phase of both plantation timber production and wastepaper utilization. Although production is extremely high, consumption is also extremely high. On a comparative international basis, domestic production is relatively balanced with domestic consumption. The United States, with less than

5 percent of the world's population, produces and consumes roughly a third of all paper and paperboard in the world.

Canada, however, with greater remaining forest resources (second only to Russia) has yet to seriously define resource limits, and most of the major new virgin capacity expansions planned or recently built in North America are located in Canada. Paper and other forest products play a larger role in the substantially resource-based Canadian economy than in the more diversified U.S. economy, and exports alone (more than 72 percent of production relative to about 11 percent in the United States) are more than double domestic consumption.

The Nordic countries (Finland, Norway, and Sweden), which contain nearly half of western Europe's forests, demonstrate a similar pattern in terms of the importance of the industry to the national economy and their prominence as leading paper exporters (73 and 89 percent of domestic paper production in Sweden and Finland respectively). However, they have long since passed the stage of reliance on natural forests and, like Germany, have extensive timber plantations as well as substantial wood imports (from Russia, Canada, and the United States). The major European producers are the technological leaders of the world industry and, arguably, operate within a stricter environmental regulatory context (when both products and facilities are considered) than exists anywhere else in the world.

All of these countries (Canada, the Nordic countries, and Germany) and Japan have substantially stronger levels of government involvement with the industry than the United States, through public/private technology development programs and sector-focused regulatory policies, although the United States, like Canada, continues to subsidize the industry more indirectly by providing cheap access to national forest resources and maintaining nineteenth-century tax structures for resource-based industries.

Russia inherited 95 percent of the former Soviet Union's forestlands and houses the world's primary wood reserve, although its liquidation is proceeding. The Russian taiga forest covers an area approximately the size of the United States, or about twice the size of the forested area of the Amazon basin. The taiga forest contains some 52 percent of the world's standing softwood. Russian pulp mills are, by North American

Table 1.2
Paper production and consumption, selected countries, 1991

	Production (1,000 tons)	Exports (1,000 tons)	Imports (1,000 tons)	Consumption (1,000 tons)	Population (million)	Per capita consumption (lbs)
World	267,819	63,478	62,333	266,674	5,283.9	101
Africa	2,982	139	1,378	4,222		
North & Central America	101,628	19,795	14,182	96,016		
Canada	18,215	13,140	1,065	6,140	26.8	458
Mexico	3,186	63	419	3,542	85.7	83
United States	79,996	6,563	11,818	85,252	252.8	674
South America	8,522	1,212	1,265	8,575		
Brazil	5,377	924	257	4,710	153.3	61
Asia	66,452	4,747	9,823	71,529		
China	20,387	807	1,989	21,569	1,151.3	37
India	2,640	10	208	2,838	859.2	7
Indonesia	1,870	399	110	1,581	181.4	17
Japan	31,958	1,318	1,361	32,001	123.8	517

Europe	74,559	36,012	34,092	72,640		
Finland	9,356	8,276	134	1,213	5.0	485
France	8,186	2,604	4,165	9,747	56.7	344
Germany	14,894	4,989	8,386	18,292	79.5	460
Italy	6,380	1,285	2,774	7,869	57.7	273
Netherlands	3,201	2,349	2,802	3,654	15.0	487
Spain	3,934	624	1,894	5,204	39.0	267
Sweden	9,191	6,695	336	2,831	8.6	658
UK	5,445	1,363	6,080	10,162	57.5	353
Oceania	3,127	595	1,010	3,542		
Australia & New Zealand	3,127	596	974	3,505	21.0	334
Former USSR	10,549	979	582	10,152	292.0	70

Sources: FAO, *Forest Products Yearbook* (1991); World Bank, *World Development Report* (1992).

and European standards, outdated, inefficient, and heavily polluting. Production is based entirely on virgin resources and concentrated around low value-added commodities. About 10 percent of total production is exported. Russia is the only major country in the world where paper and paperboard consumption has fallen in recent years.[42]

Japan, like all the major northern producers except Russia, has high levels of consumption, but is unique in having very limited timber resources (the entire country is approximately the size of California, but the population density is twelve times that of the United States). The Japanese paper industry is dependent on wood imports (from the United States, Canada, Australia, and southeast Asia) for more than 60 percent of its total wood supply and on both domestic and imported wastepaper supplies. It is the only major producer whose fiber supply comes primarily from outside the country. Less than 3 percent of its paper and paperboard production is exported.[43]

The Chinese paper industry differs substantially in almost every respect from other major producers. Nearly 90 percent of all pulp is produced from nonwood fiber sources (primarily straw), reflecting the scarcity of timber and the abundance of agricultural materials. Mills are typically a fraction of the size of western mills, with most estimated to operate at levels below 10,000 tpy (less than 30 tpd). Machinery is often outdated and originally designed for wood pulping. There is limited chemical recovery, minimal environmental regulatory oversight, and severe pollution. Paper and paperboard are generally of low quality and produced for the domestic market (only about 1 percent is exported). The per capita consumption level is among the lowest in the world.[44]

Both Japan and western European producers have considerably higher levels of wastepaper use than the U.S. By 1993, for example, wastepaper accounted for 53 percent of the fiber sources used by Japanese and German paper producers (the second and fifth largest paper producing countries), while in the United States wastepaper accounted for 31 percent of the fiber sources used in paper production.[45]

The link between paper consumption and the level of economic development within a country is well established. As a representative 1980s analysis of the global outlook for the U.S. paper industry observed, "So pervasive are the uses of paper in the business of economic growth that gross national product projections are widely used in the paper industry as a leading predictor of paper demand. As goes GNP growth, so go sales of paper."[46] However, as the following figures indicate, it is clearly past time to decouple the two and to reconsider the passivity with which we have accepted the recent levels of growth in domestic consumption.

In 1991, annual per capita paper and paperboard consumption in the United States was the highest in the world at 674 lb. (see table 1.2).[47] The only close competitor to the United States in per capita consumption was Sweden (658 lb.). Japan consumed almost a quarter less paper per capita, Canada nearly a third less, and the U.K. and France about half as much per capita. The European countries that are also major paper producers consumed, on average, about 415 lb. per capita, or nearly 40 percent less than U.S. consumers. Japan and at least six European countries had higher per capita GNP but significantly lower per capita paper and paperboard consumption.

Although this analysis is focused primarily on resource use and production, and relatively less on end-products consumption, in some respects these consumption trends are among the most import numbers in this book. Growth in domestic paper consumption in the past two decades is summarized by major grade in table 1.3 and surely qualifies as astonishing. In the 1980s alone domestic consumption increased, on average, about 3 percent annually, for a total increase of nearly 30 percent in ten years (relative to a 26 percent increase in production during the same period). The increase in total consumption between 1970 and 1990 was about 56 percent. Needless to say, with wastepaper comprising the largest single chunk of the municipal solid waste stream (between 30 and 40 percent in the past decade), the growth in paper consumption in the last two decades was not unrelated to the emergence of a solid waste management crisis in the 1980s. Franklin Associates, on behalf of the AFPA, projects a more moderate rate of growth in consumption (comparable to the 1970s) of about 19 percent in the 1990s. Although modest relative to the preceding decade, this rate of growth nevertheless represents a projected increase in consumption of nearly 14 million tons per year.[48]

Excluding consumption of packaging papers (distinct from paperboard) and of building-grade paper and board, which have been

Table 1.3
Paper consumption by grade, 1970–2000

	Total consumption (million tons)							
	Newsprint	Printing and writing	Packaging and converting	Tissue	Paperboard	Building paper and board	Total paper and board	U.S. population (resident) (1,000s)
1970	9.8	11.0	5.1	3.7	23.5	3.0	56.1	203,984
1971	9.8	11.3	5.2	3.8	23.9	3.5	57.5	206,827
1972	10.3	12.3	5.5	4.0	25.9	3.7	61.6	209,284
1973	10.9	13.5	5.6	4.0	27.6	3.6	65.2	211,357
1974	11.0	13.4	5.6	4.1	26.8	3.3	64.2	213,342
1975	9.0	10.9	4.3	3.9	23.3	3.0	54.6	215,465
1976	9.9	13.3	5.2	4.2	26.7	3.4	62.6	217,563
1977	10.1	14.2	5.6	4.3	27.2	3.5	64.8	219,760
1978	10.8	15.2	5.8	4.2	28.3	3.7	68.1	222,095
1979	11.2	16.1	5.7	4.5	29.3	3.6	70.3	224,567
1980	11.4	16.1	5.2	4.3	27.7	3.0	67.8	227,225

1981	11.7	16.1	5.2	4.5	28.7	2.5	68.7	229,466
1982	10.9	16.2	5.0	4.4	26.5	2.0	65.1	231,664
1983	11.6	18.5	5.3	4.8	29.3	2.2	71.7	233,792
1984	12.7	20.5	5.4	4.9	31.5	2.2	77.3	235,825
1985	12.8	20.6	5.1	5.0	30.6	2.0	76.2	237,924
1986	13.0	22.1	5.1	5.1	32.5	1.8	79.7	240,412
1987	13.7	23.4	5.0	5.3	34.6	1.8	83.9	242,289
1988	13.7	24.8	5.0	5.5	35.3	1.8	86.1	244,499
1989	13.1	24.4	5.0	5.7	35.9	1.7	85.8	246,819
1990	13.3	25.5	4.7	5.9	36.5	1.7	87.6	249,391
1991	12.5	24.7	4.6	5.7	36.8	1.7	86.0	252,160
1992	12.6	26.0	4.8	5.8	38.6	1.7	89.5	255,082
2000	14.7	30.9	5.1	6.8	44.7	1.6	103.8	274,900

Sources: Franklin Associates, Ltd., *The Outlook for Paper Recovery to the Year 2000* (1993) and *Paper Recycling: The View to 1995* (1990); U.S. Department of Commerce, *Current Population Reports* (1993).

displaced by the increased use of plastics and other synthetic materials, per capita consumption in all other grades has increased by between 10 and 90 percent in the past two decades (1970–90) and by about 30 percent overall. This means that on Earth Day 1990 we were using about 151 lb. more paper and paperboard per person than we were on Earth Day 1970. In the mid-1990s we are probably using 10–20 lbs more paper per capita than in 1990.

The largest consumption increase has been in printing and writing grade papers: a 44 percent increase in per capita consumption in the 1980s alone. As Claudia Thompson explains:

> The rapidly expanding use of computers, copiers, and facsimile machines has played a significant role in these increases. Futuristic predictions of "the paperless office," which glorified the wonder of these new technologies, have proven naive. They failed to account for the propensity of people to reproduce many more documents at more frequent intervals.[49]

It is, of course, not only "the propensity of people" but also of properties intrinsic to the technologies that cause them to be stimulants of paper consumption. Broader factors driving the increased consumption of printing and writing and other "cultural" papers such as newsprint include a growing segment of service and administration in the national economy and developments favoring print media over other media as a vehicle for advertising, with an increase in direct mail and catalogues. Anyone who has ever unpacked a computer or other consumer appliance or toured the backyard of the local supermarket or other retailers can attest to the significant role that consumer products packaging and shipping materials have played in the 25 percent increase in per capita paperboard consumption during the last decade.

To the paper industry these numbers must be very encouraging; from an environmental perspective, however, they are profoundly disturbing. A 1994 analysis by the U.S. Forest Service, for example, came to the following conclusions in response to the question, "Will increased paper recovery (for recycling and export) either eliminate or reduce the disposal burden associated with wastepaper in the United States?"

> No. The gross wastepaper disposal burden of the United States will remain in the range of 45 to 55 million metric tonnes per year well into the 21st century. . . . [A]lthough recovery of paper for recycling and export will continue to grow,

domestic demand and consumption of paper and paperboard products will also continue to grow in the future. Thus the United States will continue to face a gross wastepaper disposal burden in the decades ahead, which will be similar in magnitude to the current disposal burden.[50]

Indeed, many of the anticipated benefits of paper recycling, including solid waste reduction, forest conservation, and pollution prevention have been lost under the burden of still-rising production and consumption levels, as will be discussed more in later chapters. The competitive forces of free markets are hardly likely to generate a reduction in the consumption of paper in the foreseeable future: at least not until all but the last few remnants of functioning forest ecosystems have been either taken off-limits, destroyed, or converted to the most productive and optimized "supertree" plantations that genetic engineering can provide, and not until consumption has met a price it can't pay in the face of physical limits to production.

At the same time, deliberate and effective strategies to reduce consumption in advance of increasingly unpleasant futures are not something that modern industrial societies and their governments have shown themselves to be particularly good at; indeed, the very concept flies in the face of free market and free trade ideology. The United States has risen to the task now and then, when the problem is extreme and incontestable, as in banning or restricting the use of a handful of extremely toxic pesticides and other chemicals, and in mandatory and voluntary rationing of consumption to support war efforts in the 1910s and 1940s. Yet despite the position of source reduction at the top of public policies for waste management, reducing or even *stalling* the growth in consumption of paper and paperboard has yet to be undertaken seriously.

Some greener (or more land use limited) European governments have lately shown more encouraging signs in terms, for example, of serious efforts to reduce packaging use through various economic sticks and carrots. However, a major industry whose markets are threatened and that is structurally maladapted for change of any sort can effectively forestall progress. In the mid-1990s, with public policy mandates for source reduction of waste (albeit relatively weak ones) in place for at least a decade in both Europe and the United States, the industry itself is only just beginning to awaken to the notion that reducing consumption of its products might become a serious public policy objective. Its response has been predictable. For example, in a recent review of various tax and regulatory schemes applicable to paper products being contemplated in Europe, one industry analyst wrote, "The pulp and paper industry should . . . realize that all such environmental concerns could one day lead to the consumption of paper being seriously questioned. The industry must mount a common defense."[51]

When one considers the degree of seriousness the solid waste disposal problem had to reach in the United States before recycling even reemerged as a national priority, as well as the high level of national mobilization it has taken to increase wastepaper recovery rates, it is clear that neither reducing paper consumption nor significantly influencing the evolution of the raw materials and production base of the paper industry are tasks for the fainthearted or naive. The industry itself is structurally disinclined to either significant or rapid change, and public policy has largely failed to appreciate (or even clearly identify) the monumental inertia of an old, stable, capital-intensive, and strongly wood-oriented industry whose products permeate every aspect of the economy and society.

Ashford and Heaton write, however, of technologies (and associated industries) in their mature phase of development: "At this point in its life cycle the technology may be subject to invasion by new ideas or disruption by external forces that would cause a reversion to an earlier stage."[52] The combined forces of the broad-based and heterogeneous American environmental movement—expressed in regulatory and public protest forms and, more lately, in a handful of economic development efforts and focused on increased recycling, pollution prevention, and energy and forest conservation—represent such major new ideas and disruptive external forces for the paper industry. If they are not all new ideas, then there are at least signs that they are more urgent and will be more enduring than in the past. If there is any doubt about the importance of these forces from the industry's perspective, consider the comments of Peter Wrist in delivering the Gunnar Nicholson Lecture to the 1992 Environmental Conference of the Technical Association of the Pulp and Paper Industry:

> Today our industry is under savage attack by groups claiming that our operations are destroying irreplaceable natural treasures; poisoning the oceans, lakes, and

> rivers; burying our cities under piles of garbage; and threatening the health of the public and the future of the world. This sudden change in public opinion is the result of a major environmentalist campaign, building upon a heightened environmental awareness amongst the public and a change in the ethics and values of affluent developed countries as they move into what is being called the "post-industrial society." The impact of this sudden change of public opinion upon our industry's markets has been dramatic, bringing with it major changes in product specifications and increased government regulation in all facets of our activities.[53]

Having awakened the industry and provoked a debate that essentially raises the question of what a twenty-first-century paper industry will look like, the challenge for policymakers and environmentalists is now to help redefine the structure of this industry. To do so they must clearly define the parameters within which it will operate, with the understanding that these parameters must cross a wide variety of connected economic, environmental, and social policy terrains.

2

Forests and Fiber Resources

Several fundamental issues underlie and complicate all discussions of forests, timber production, and wood utilization. One issue is the extensive integration of the forest products industries introduced in the preceding chapter—a subject pursued here in terms of how this integration is organized around materials flows. A second issue involves the multiple effects of trade and export policies and practices surrounding not only logs, lumber, and paper, but also the variety of less obvious intermediate forms in which wood-based commodities such as pulp, wood chips, and even wastepaper are traded—a subject considered here and in subsequent chapters. A third issue strongly related to forests and timber production, but rarely directly associated with their analysis, concerns the availability of alternative fiber resources, including wastepaper, annual fiber crops, and agricultural residues. Possibilities related to such alternatives pose a serious challenge to many of the core assumptions long embedded in forest and timber policies.

The final and most important issue, however, upon which all forest and timber analyses are built, is a fundamental dichotomy in the ways in which the subjects of forests and timber have come to be viewed. The modern business of cutting down trees cannot be unequivocally described as either a renewable *agricultural* industry or as an *extractive* industry (like mining, or oil production) where the original resource is not considered renewable. As historically and currently practiced, timber production has characteristics of both. In essential respects the distinction between the two is well described by the difference (as the saying goes) between the forest and the trees. At one extreme, for example, the large-scale clear-cutting of old growth and other late successional forests is clearly an essentially irreversible extractive process. No one has proposed that we can deliberately recreate them in all their complexity once they are gone, and the time scales for natural regeneration to a mature forest ecosystem are in any case measured in centuries. The process of natural regeneration further implies an absence or subtlety of human presence over such significant periods of time that its prospects are difficult to entertain for the future.

At the same time, certain fast-growing types of trees like pines and eucalyptus are now being grown in plantations on rotations currently estimated to be feasible in the range of about 15–25 years for southern U.S. pines and reportedly as short as 5–7 years in hybrid eucalyptus plantations in the southern hemisphere. Yet, although such plantations have much in common with conventional agricultural crops, they have very little—except trees—in common with forests. Forests can be defined in various ways, but a common use of the term describes an ecosystem type, much like the terms *desert, wetlands* or *prairie* refer to other basic ecosystem types. In this view, forests are natural biological systems made up of a diverse array of flora and fauna interacting in complex ways with each other and with soil, air and water. Trees are simply the most obvious part.

In some respects, intensively managed timber plantations, especially monocultures grown in short rotations, can be viewed as both analogue and sequel to the clear-cutting of the virgin forest, to the extent that neither clear-cutting nor plantations allow for the reproduction of complexity and function inherent in a natural forest ecosystem. Although timber may be renewable, only some forms of timber production allow forest biodiversity to be renewed and sustained. The traditional practice of timber production in the United States, often referred to as *timber mining,* has been slowly and unevenly transitioning to the practice of *sustainable timber production* or *sustainable yield* in recent decades, augmented by requirements for multiple-use management of public forests. More lately these concepts have begun to give way to new attempts to define *sustainable forest management.*

If the wholesale land-use transformation associated with both large-scale clear-cutting and intensive timber cropping can be said to occupy one extreme of a continuum of forest management options, then both protected wilderness and the low-impact forest use practices of many traditional societies occupy the other. Inhabiting the extensive middle ground are such practices as selective clearcutting, soft forestry, and various other forest management approaches that more carefully harvest wood from forests by using comparatively low-impact logging techniques, plant seedlings, and generally attempt to preserve essential characteristics of the native forest. In parallel, a handful of more sophisticated efforts in plantation management seek to mimic aspects of natural forests through the cultivation of multiple species in mixed-age stands and longer rotations.

Although the value of timber as a commodity can be measured, forests cannot be similarly deconstructed into simple economic terms despite the best efforts of diligent resource economists. Increasingly in recent years, there have been theoretical attempts to quantify present and future values of forest biodiversity, the value of various (poorly understood) "ecosystem services" such as climate and watershed regulation, as well as the amenity (recreational) and even religious and cultural values of forests.[1] Not surprisingly, such attempts have failed to do more than suggest the vast dimensions of our ignorance. Even the depletion of the simple asset value of the timber stock in forests (the present value of potential income from the harvest of standing trees) is not reflected in the national economic accounting system.[2] Although timber production is treated as a subject of such vital importance that it is tantamount to national security, the condition and value of the underlying forest resource has been substantially ignored in both economic and ecological terms. Catherine Caufield points out, for example, that the first comprehensive ecological study of the great Pacific forest of North America, "a forty-eight page report," was published as recently as 1981.[3]

The debates over forests and timber and over associated public and industry policies have, of course, become polarized to the point of schizophrenia ("Owls don't pay taxes!"). Much of the rhetoric, however, can be understood to turn largely on whether one views (or at least portrays) the modern business of timber production as an extractive industry or as a renewable industry. One can quickly turn, for example, from a literature of heartbreaking images and vehement denunciations of the rape of the forests to mainstream forest industry publications that cheerfully announce "we're planting faster than we're cutting." The problem is that what they have been planting often bears only a trivial relationship to what they have eliminated. The jargon of agriculture, with its overtones of cyclical renewability, is pervasive in discussions of timber resources, and the U.S. Forest Service is in fact housed within the Department of Agriculture. This masks, however, what has substantially been and continues to be an unrenewable, essentially extractive process of destruction—at best a simplification and shallow mimicry of the structural, functional, and genetic diversity of healthy forest ecosystems.

The primary statistical data sources on timber stock, total production, growth rates, and so on are suffused with this oversimplified agricultural rhetoric and perspective. They fail not only to reflect the simple asset value of resource (timber) depletion, but also (single-mindedly focused on trees) to formally distinguish any other impacts of production. As Perry Hagenstein dryly observes, "[M]easures of overall biological productivity in relation to the land's potential, and direct measures of [forest] condition in relation to uses other than timber growing are generally lacking."[4] Thus, with few other measurements available in the aggregate statistics, a million cubic feet of growth in timber stock is equal to a million cubic feet logged, regardless of whether the growth was in intensively managed timber plantations and the logging was in a mature forest. Sustainable timber production continues in many ways to be used as the proxy for sustainable forest management, while debates over how to define and certify the latter continue to develop.[5] These are just some of the handicaps and pitfalls that underlie the following quantitative characterizations of the impact on forest resources by the forest products industry in general and by the paper industry in particular.

Timber Production and Utilization

The United States is by far the largest producer and consumer of timber in the world. In 1991, domestic *roundwood* production (defined as all "wood in the rough" felled or otherwise harvested, including recovered roots and other logging residues, but primarily referring to logs) amounted to nearly 18 billion cubic feet (bcf) or nearly 15 percent of estimated global roundwood production. The United States was followed by the former Soviet Union (12.6 bcf); Europe (11.8 bcf, led by Sweden at 1.8 bcf); China, India, and Indonesia (each about 9.9 bcf); Brazil (9.4 bcf); and Canada (6.2 bcf). Together, the United States and Canada accounted for almost 20 percent of global roundwood production (see table 2.1).[6]

At a world level, more than half the trees cut down each year are used for fuel, although the divergence between the industrialized and the developing world is well known and extreme. Among the developed nations, fuel and charcoal account for only 16 percent of all roundwood uses, whereas in the developing world they account for more than 80 percent and well in excess of 90 percent in the poorest countries.

Nonfuel primary uses of roundwood are characterized as *industrial roundwood* consumption. The primary consumers of industrial roundwood, as described in the preceding chapter, are lumber manufacturers, plywood and veneer manufacturers, the pulp and paper industry, and miscellaneous primary wood products manufacturers (e.g., manufacturers of telephone poles). Industrial roundwood is thus categorized as *sawlogs and veneer logs, pulpwood,* and *other industrial roundwood.*

In 1991, U.S. roundwood exports consumed 6 percent of domestic roundwood production, having risen steadily since the early 1950s from about 10 million cubic feet to almost 1.1 billion cubic feet in 1991. Imports of roundwood were minor, and net exports have averaged 4–6 percent of roundwood production in recent decades. The United States has the questionable distinction of being the largest exporter of unprocessed wood in the world. Accounting for about 15 percent of world roundwood production in 1991, the United States supplied nearly a quarter of all roundwood exports. Its total roundwood exports were roughly double those of the nearest competitors: Malaysia and the former Soviet Union. For the past several decades the largest part of these roundwood exports (which include both logs and wood chips) have consisted of extremely high-quality softwood logs from old-growth (more than two hundred years old) forests in the Pacific Northwest, destined for Japan (about 60 percent) and other Asian countries.[7]

This has rightly been a subject of enormous controversy. In 1973, log exports from federal lands west of the one-hundredth meridian were banned, as were log exports from lands owned by the State of Oregon.

Table 2.1
Roundwood and sawnwood production, selected countries, 1991 (million cubic feet)

	Roundwood						Sawnwood	
	Fuelwood and charcoal	Sawlogs and veneer logs	Pulpwood	Other industrial roundwood	Total roundwood	Net roundwood exports	Total sawnwood	Net sawnwood exports
World	64,631	33,043	15,161	8,274	121,108	100.0	(158)	16,156
Africa	16,538	826	277	976	18,617	15.4	139	298
North and Central America	5,085	13,479	6,861	615	26,040	21.5	909	5,651
Canada	241	4,358	1,607	81	6,288	5.2	(78)	1,838
Mexico	560	200	69	5	834	0.7	3	95
United States	3,034	8,800	5,181	494	17,509	14.5	982	3,669
South America	8,562	1,972	1,388	277	12,198	10.1	239	910
Brazil	6,715	1,341	1,084	205	9,345	7.7	(3)	607
Asia	29,457	5,907	754	2,239	38,356	31.7	(1,800)	3,561
Bangladesh	1,090	16	2	13	1,121	0.9	(4)	3
China	6,789	1,600	279	1,303	9,970	8.2	(302)	725

India	9,015	648	43	175	9,881	8.2	(44)	617
Indonesia	5,074	925	7	3,875	9,881	8.2	0	323
Iran	87	11	0	6,010	6,109	5.0	(2)	6
Japan	12	623	344	20	998	0.8	(1,753)	998
Malaysia	316	1,380	22	25	1,742	1.4	677	315
Pakistan	874	72	0	14	960	0.8	(1)	54
Philippines	1,210	55	12	91	1,368	1.1	(12)	26
Thailand	1,238	8	0	93	1,340	1.1	(61)	33
Viet Nam	872	106	0	65	1,043	0.9	14	31
Europe	1,817	5,336	3,936	758	11,847	9.8	(549)	2,886
Finland	105	506	570	23	1,204	1.0	(185)	211
France	369	823	369	19	1,580	1.3	113	388
Germany	159	765	484	177	1,585	1.3	245	535
Sweden	156	798	867	5	1,827	1.5	(136)	407
Oceania	309	675	476	40	1,499	1.2	392	184
Australia & New Zealand	104	292	470	310	1,176	1.0	317	175
Former USSR	2,864	4,849	1,469	3,369	12,551	10.4	561	2,666

Source: FAO, *Forest Products Yearbook* (1991).

However, the conflict that led to efforts to eliminate these exports was not, in the 1950s and 1960s, primarily environmental, but rather focused on preempted economic opportunities. A valuable resource that underlay one of the most significant basic materials flows in the economy was being exported at the lowest possible link of the value-added chain. Within the most highly developed national economy in the world, the Pacific Northwest region has for decades emulated the behavior of the least-developed economies through significant exports of its virgin resource base.

That a large part of these roundwood exports came from a publicly owned resource simply added to the insult. After the federal ban, old-growth log exports continued to originate from industry-owned lands in Oregon and Washington, and state-managed lands in Washington (as well as from British Columbia). By the early 1980s, however, the U.S. Forest Service could accurately predict that most of the high-quality (old-growth) timber on these state- and industry-managed lands would have been harvested by the 1990s.[8] After annual growth in roundwood exports that averaged above 6 percent between the mid-1980s and early 1990s, it now appears that the wood export tide has begun to slacken due to the depletion of the virgin forest resource. Between 1991 and 1993 total wood exports (including both roundwood and lumber) from the United States declined by 18 percent, and the source of domestic wood exports has increasingly shifted from the Pacific to the South Atlantic states.

The subject is important for a broader perspective on timber production, wood use, and the paper industry because it highlights the importance of where, on the chain that stretches from the resource base to a diverse array of wood-based end products, exports occur. The lower on the chain of value-added manufacturing—the closer to the resource base—that exports fall, the fewer the employment and other benefits relative to the resource depletion and environmental costs. Additionally, because the paper industry is so strongly integrated with the forest products industries as a whole, changes that may appear beneficial in materials use and trade at the higher-value levels of manufacturing, must also be examined in terms of their effects on materials use and trade at the level of primary and intermediate wood-based commodities.

To a large extent, the undesirable economic and environmental aspects of exporting roundwood are duplicated in exports of lumber, woodpulp, and wastepaper. In 1993, for example, the average value of wood exports (roundwood, chips, and lumber) was \$275/ton, woodpulp was \$375/ton, and wastepaper was \$82/ton, whereas the value of paper and paperboard exports averaged \$508/ton.[9]

As with both roundwood and woodpulp exports, U.S. lumber exports have been steadily rising, accounting for more than 9 percent of total lumber production in 1991. Canada is the largest exporter of lumber in the world, and the United States is the second-largest. Like roundwood exports, most lumber exports in recent years—more than 70 percent—have originated from the Pacific Northwest. However, as with paper, the United States remains a net importer of lumber due to its huge domestic markets. Whereas paper and paperboard imports have declined to about 4 percent of domestic consumption, lumber imports (mostly from Canada) continue to account for more than 15 percent of domestic lumber consumption. Thus, the United States is a net exporter of unfinished commodities (roundwood, woodpulp, wastepaper), and a net importer of lumber and paper.

Unlike paper consumption levels, average per capita lumber consumption in the second half of this century has been *lower* than in the first half, although particular decades have been characterized by sharp fluctuations. During the early decades of the century lumber consumption ranged from as high as 362 board-feet per capita (1923) to a low of 102 (1932) during the depression. Since the 1950s lumber consumption has generally ranged between about 200 and 240 board-feet per capita; at the lower end of the range through the 1970s and rising to the higher end during the 1980s.[10] In the same period (about 1920 through 1990) per capita paper consumption increased nearly fivefold and continues to climb. Pulp and paper production have thus, particularly in the last several decades, played a disproportionate role in the increased demand on wood resources in the United States.

In 1991, total domestic roundwood consumption was 16.5 bcf, apportioned among various uses as follows: sawlogs and veneer logs (50 percent), pulpwood (29 percent), fuel (19 percent), other industrial (3 percent). Lumber manufacturers alone consume more than 75 percent of sawlogs and veneer logs, or about 38 percent of total domestic roundwood consumption and 50 percent of industrial roundwood consumption. The paper industry consumed more than 4.8 bcf of pulpwood in 1991, or about 36 percent of industrial roundwood.[11] Pulp production thus constitutes the second-largest source of *primary* demand on domestic forest resources. Figure 2.1 illustrates the shifting importance of these various wood uses since 1950. In that year, for example, pulpwood represented less than 15 percent of primary domestic roundwood uses, relative to nearly 30 percent at present.

What these figures obscure, however, and what is generally less well understood and documented, is the significant integration between pulp producers and lumber and other wood products producers around the use of *wood residues:* shavings, slabs, chips and sawdust. This secondary materials flow in fact substantially supports the corporate organization

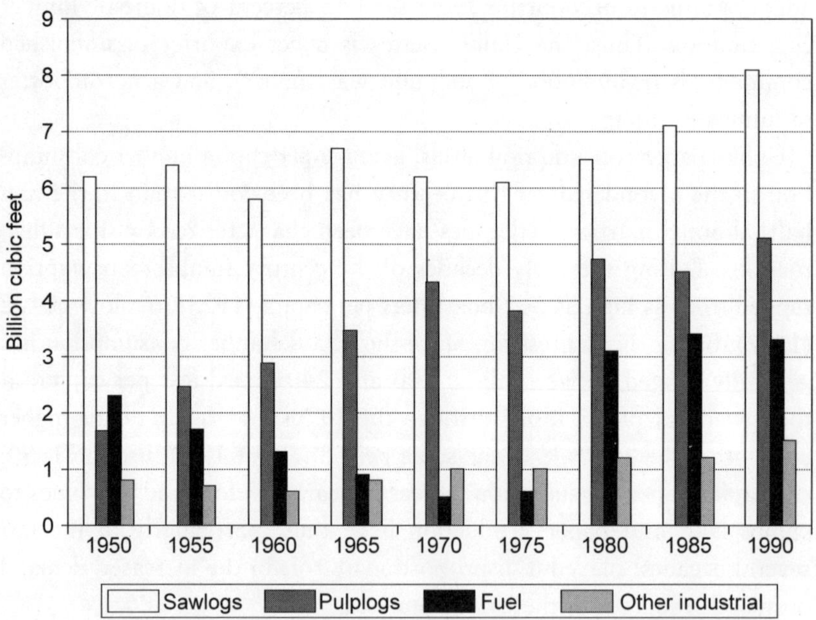

Figure 2.1
Roundwood production, 1950–1990
Source: Council on Environmental Quality, *Environmental Quality 1993.*

discussed in the preceding chapter. Because there are only a certain number of solid rectilinear objects (lumber) that can be produced from cylindrical objects (logs), the yield, or proportion of lumber produced to roundwood consumed, averages only about 38 percent (softwood) to 49 percent (hardwood). In other words, *some 50 to 60 percent of the wood that enters a sawmill emerges in the form of wood residues.* Similarly, the yield of plywood and veneer products ranges from about 45 to 57 percent of the incoming roundwood, leaving the remainder as residues.[12] On balance, slightly more than half of the roundwood (sawlogs and veneer logs) that enters sawmills or plywood and veneer mills emerges as wood residues, and these residues have become a valuable commodity in their own right. In fact, it is more accurate to view wood residue as a valuable *co-product* than as a *by-product* of lumber and other wood products manufacturing.

Figure 2.2 outlines the flow of both roundwood and wood residues to all uses in 1986—the latest year for which comprehensive estimates on the use of wood residues by sector is available.[13] For that year the U.S. Forest Service calculated that the use of wood residues from lumber, plywood, and veneer mills was apportioned as follows: pulp (55 percent); fuel (28 percent); particleboard, fiberboard, and miscellaneous industries (11 percent); and export (3 percent). Only about 3 percent of wood residues went unused, producing an overall efficiency of roundwood use approaching 100 percent.

When one views the consumption of wood by pulp manufacturers in terms of both primary *and* secondary consumption (i.e., both roundwood and wood residues), the real significance of wood use by the paper industry emerges. Total (primary and secondary) wood consumption by the paper industry in 1986 was more than 7.5 bcf, or about 43 percent of domestic roundwood consumption. By 1991 this was close to 8 bcf, approaching 50 percent of total domestic roundwood consumption. *Within the context of total (primary and secondary) wood consumption, the U.S. paper industry is the single largest consumer of wood in the country and the largest industrial consumer of wood in the world.* Of all the timber harvested from domestic timberlands each year, more ends up in paper than in buildings, furniture, and all other wood products combined.

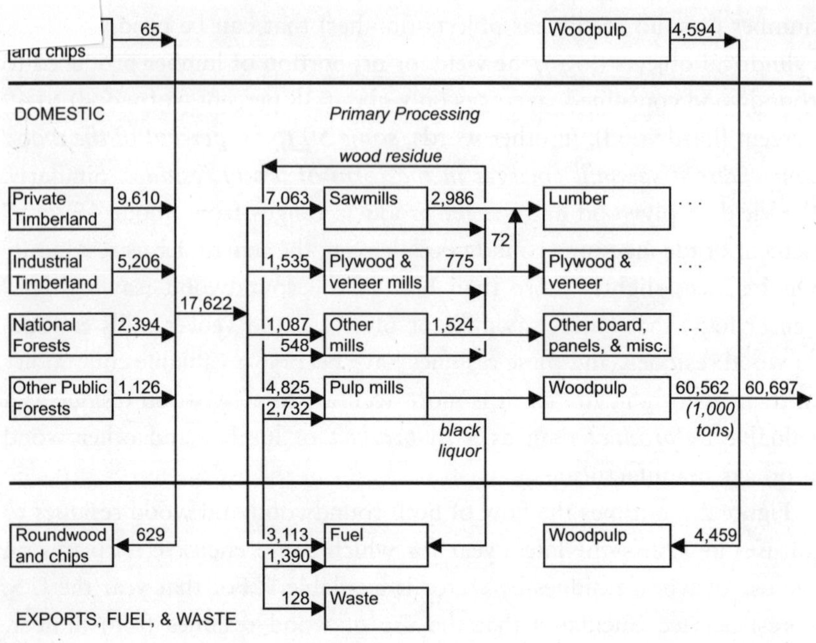

Figure 2.2
Wood flow, 1986 (million cubic feet)
Source: U.S. Dept. of Agriculture, Forest Service, *An Analysis of the Timber Situation in the United States: 1989–2040* (1990).

In 1991 the total fiber used in domestic paper production was derived from pulpwood (45 percent), wood residues (27 percent), wastepaper (28 percent), and nonwood fibers (negligible at less than 1 percent).[14] Although wood residue should be viewed as a valuable co-product of lumber production, it often continues to be cast as a waste by-product, and thus its use by the paper industry is seen as somewhat akin to the use of wastepaper. Indeed, in the spring of 1990 controversy arose when the EPA ruled that sawdust used in the manufacture of pulp at two mills in Maine could be classified as a "recovered material" and used in meeting minimum content standards for the EPA's recycled paper procurement guidelines.[15]

The controversy, however, highlighted not only the obvious recycling implications of determining whether wood residues are a primary raw material or a recovered waste material for the pulp and paper industry, but also the difficulty of determining the extent to which the industry's consumption of wood residues produced by other manufacturing sectors creates a demand on forest resources in its own right. Considering that wood residue constitutes some 40 percent of the wood supply to a more than $130 billion industry, its use cannot be described simply as a passive efficiency measure. Within the context of integrated wood-based companies in particular, paper production, which produces a higher value than lumber production relative to equivalent wood consumption levels, can be viewed in some respects as a form of subsidy or incentive for lumber production. Similarly, various forms of tax subsidy to resource-based industries, including timber production, allow integrated companies to shift their overall taxable income back into less-taxed raw materials operations, providing for what amounts to a two-way subsidy: paper production may subsidize lumber production by creating a strong market for wood residues, while the tax advantages associated with timber production can be extended to high-revenue paper production.[16]

In all countries with significant forest products industries, both investment and disinvestment in timber, lumber, and woodpulp production and trade are strongly linked. Pulp mills, along with residue-based panel mills are the natural partners of sawmills and of plywood and veneer mills, tied together by the production and consumption of wood residue. With the rapid rise of industrial timber plantations, however, the strength of the relationship may be moderating somewhat. The option arises of whether to invest in the extra time needed to bring timber crops to an age and size that can support integrated lumber and pulp production or to harvest young trees on short rotations dedicated to pulp production. Nevertheless, the integration of primary wood-processing industries around wood residues is a very strong force and a defining characteristic of the forest products industries. The relative invisibility of the wood residue flow has been a major factor in the pervasive tendency to understate the true magnitude of the demand placed on forest resources by the paper industry. Typically, only primary demand rather than total wood use is considered.

The Distribution of Production

In her classic 1990 essay entitled "The Ancient Forest," Catherine Caufield wrote:

> The loggers started in the great hardwood and white-pine forests of the Northeast. By the time of the Civil War, those forests had been largely exhausted. Logging then moved to the pineries, the cypress swamps, and the live-oak stands of the South and to the pine forests of the Great Lakes region. It took only fifty years to deplete the latter. One writer of the time described logging as "the great nomad among American industries, driving from one virgin forest to another, like a threshing machine from one ripe wheat field to another." . . . In the West, the loggers came up against their last frontier—the most magnificent forest on the continent and the greatest conifer forest on earth.[17]

Many forest historians have written of the march of this threshing machine across the continent and of the denial that marked its progress and the wastelands it left in its wake. With more than 90 percent of the original Pacific forest of the United States logged in the past forty years, the industry has indeed exhausted all but remnants of its last great domestic frontier. Its embrace of the rhetoric of sustainability in recent years—even as it insists on access to those remnants—grates on the sensibilities of those who know anything of its history and its methods. The industry now asserts (speaking on behalf of the society) that "we are at a point of fundamental change in the way society values and manages its forest resources." More to the point, however, "The traditional production forestry concept . . . may no longer be viable."[18]

In the four decades between 1950 and 1990, the annual timber harvest in the United States increased by 66 percent (7.2 bcf). As with paper production and consumption, however, most of this growth occurred between 1970 and 1990, when the annual harvest increased by 55 percent (6.4 bcf), relative to a population growth of 22 percent.[19] It is worth noting that in these two decades that saw the timber harvest increase by more than half, employment in logging and sawmills declined by about 10 percent (24,200 jobs)—a not insignificant basis for the argument that technological advances (labor productivity growth) in production have been the single-most important source of job loss in timber industries.[20]

Despite the seemingly obvious implications of these ballooning timber harvests, the continuously resurging fear of timber shortage is, astonishingly, almost never attributed to rising production and consumption. Indeed, this form of denial has long been characteristic in the history of forest depletion. William Cronon, for example, in describing the rise and fall of the lumber trade in Chicago (based on the Great Lakes pine forests mentioned above) in the late nineteenth century, felt need to underscore this point: "Chicago lost its lumber trade because the forest was finally exhausted." More commonly, emerging and prospective regional or national timber shortages are attributed to environmental regulatory restraints, or to the need for more sophisticated forest management technologies. In this view, the industrial demand—however exorbitant it may be—is taken as a given, and it is the resource or the efficiency with which the resource is exploited that is the focus of attention.

The resource base that supports these levels of production can be alternately viewed as the land or the total volume of standing trees on the land. Of the 2.3 billion acres of land in the United States, approximately one-third, or about 730 million acres, is forested. This represents a loss of more than 300 million acres of forestland since colonial times, most of which was converted to agricultural use in the nineteenth century. Much of the remaining forestlands have been cut over and more or less naturally regenerated to second- or even third-growth forests, although many of these have only trace characteristics in common with the virgin forests they succeeded. This slow process of relatively natural regeneration is now infeasible. There are no new virgin frontiers (at least in the United States) to exploit while the cut-over forest regenerates itself, and the demand has mushroomed to historically unimaginable levels.

Forestland is typically defined as one or more acres of land that is at least 10 percent tree covered or capable of is being regenerated to tree cover. *Timberland* is land capable of growing timber at a rate of at least 20 cubic feet per acre per year. Approximately 71 percent (520 million acres) of U.S. forestland is timberland, of which about 6 percent (35 million acres) is *reserved timberland,* or land withdrawn from timber harvesting by administrative regulation or statute, such as federal wilderness designation.

Commercial timberland includes the remaining 483 million acres of U.S. timberland available for commercial timber harvest and is classified into the following categories of ownership: (1) *nonindustrial private* (farmers and other private landowners who are not part of the forest products industries); (2) *industrial* (forest products industries); (3) *national forest;* and (4) *other public* (primarily state, county, municipal, and Indian timberlands, but also including federal Bureau of Land Management timberlands and miscellaneous other federal lands). Table 2.2 shows the distribution of total commercial timberland among these classes of ownership for the period from 1952 through 1987, with U.S. Forest Service projections to the year 2000.[21]

The distribution of timberland by ownership has been relatively constant since the middle of this century; however, total commercial timberland has declined by more than 6 percent in the last four decades as land has been converted for development. The bulk of commercial timberland (57 percent) is concentrated in the nonindustrial private ownership class, although forest industries have substantial acreages under long-term lease from private owners.[22] The forest industries, which currently own 15 percent of the total, are the only group that increased its direct holdings (by nearly 12 million acres) since the 1950s, although the pace of industry acquisitions has fluctuated in recent years. Ronald Slinn argues that the dramatic rate of acquisitions and mergers in the pulp and paper industry in the 1970s and 1980s was driven in part by a 1973 Forest Service report that projected a timber shortage as early as the year 2000. This led financial analysts to downgrade the equities of forest products companies that were not substantially self-sufficient in wood, stimulating a forestland buying spree. In addition to mergers and acquisitions of forest products companies with substantial forest holdings by such notorious corporate raiders as Sir James Goldsmith, nearly 3 million acres of new industrial forestland, mostly in the South, was acquired between 1970 and 1987. However, a few years later, when the projected timber shortage was subsequently revised and deferred until further into the next century, companies reexamined their timberland ownership policies. They found that the rate of return on timberland ownership often fell below their desired rates of return and that, "depending to some degree on the accounting procedure, a cord of roundwood from company holdings is usually more expensive than a cord from any other source." Nevertheless, the security of owning timberlands provided an important offsetting advantage.[23] By the late 1980s, however, as will be discussed in more detail below, timber shortages were once again being projected.

National Forests account for about 18 percent of total commercial timberland, and other publicly owned timberlands account for about 11 percent. Within the latter, ownership is apportioned among states (52 percent), the Federal Bureau of Land Management (23 percent), counties and municipalities (14 percent), and Indian lands (11 percent). Overall, federal ownership including National Forests and BLM land is about 22 percent of total commercial timberland.

The distribution of ownership across different regions of the country varies dramatically (see table 2.2). The major distinction is that private land (both industrial and nonindustrial) is strongly concentrated in the South and North, whereas public lands are strongly concentrated in the West. The Pacific and Rocky Mountain states, for example, account for 77 percent of commercial National Forest timberland, whereas the North accounts for only 11 percent and the South for 12 percent. Industrial timberland, on the other hand, is distributed between the South (52 percent), North (26 percent), and West (21 percent). Nonindustrial private timberland is similarly concentrated in the South (50 percent) and North (40 percent).

The volume of standing trees on commercial timberland is known as the *timber inventory,* and includes both live trees and undecomposed dead trees. More commonly discussed is the *growing stock inventory,* which includes only live trees with commercial potential. In general, public forests have larger volumes of growing stock relative to acreage than do private forests, primarily because they must usually be managed for multiple uses. To preserve minimum recreational, hunting, and habitat uses of the forest usually requires that at least a few old (large volume) trees remain standing. By the late 1980s, for example, public forests accounted for only 28 percent of commercial timberland, but almost 40 percent of growing stock volume. Private forests had 72 percent of the land and only 60 percent of the timber volume.[24]

Changes in the timber inventory are a function of the rate of growth of live trees (which varies dramatically by age of tree, species, region, and level of management) minus the harvest. A sustainable rate of timber production occurs (within a particular area) when the net change in growing stock volume over time is zero or greater. It is worth reemphasizing that there is no equivalent "forest inventory" that reflects the

Table 2.2
Ownership of commercial timberland, 1952–2000 (million acres)

	1952	1962	1970	1977	1987	2000*
Northeast	73.0	77.9	78.0	78.5	80.1	80.1
All public	7.3	7.5	7.8	8.2	9.8	10.0
Forest industry	10.1	10.1	12.2	12.8	12.6	12.5
Other private	55.6	60.3	58.0	57.5	57.7	57.6
North Central	81.3	78.8	76.4	74.9	74.6	74.3
All public	23.0	21.9	21.7	21.2	21.2	21.2
Forest industry	3.6	3.6	5.0	4.7	4.4	4.4
Other private	54.7	53.3	49.7	49.0	49.0	48.7
Total North	154.3	156.7	154.4	153.4	154.7	154.4
All public	30.3	29.4	29.5	29.4	31.0	31.2
Forest industry	13.7	13.7	17.2	17.5	17.0	16.9
Other private	110.3	113.6	107.7	106.5	106.7	106.3
Southeast	89.0	91.0	90.0	87.8	84.6	82.2
All public	8.0	8.3	8.2	8.5	8.8	8.8
Forest industry	13.9	14.8	15.6	15.3	16.8	17.0
Other private	67.1	67.9	66.2	64.0	59.0	56.4
South Central	115.5	117.6	113.3	110.5	110.7	109.0
All public	9.7	9.7	10.2	10.1	10.9	11.2
Forest industry	17.9	18.8	20.3	21.5	21.4	21.7
Other private	87.9	89.1	82.8	78.9	78.4	76.1
Total South	204.5	208.6	203.3	198.3	195.3	191.2
All public	17.7	18.0	18.4	18.6	19.7	20.0
Forest industry	31.8	33.6	35.9	36.8	38.2	38.7
Other private	155.0	157.0	149.0	142.9	137.4	132.5
Pacific NW	44.8	43.5	43.7	42.1	38.9	38.1
All public	25.7	25.3	25.0	24.3	22.4	21.6
Forest industry	9.1	9.4	9.6	9.9	9.7	10.0
Other private	10.2	9.8	9.1	7.9	6.9	6.4
Alaska	20.4	20.1	20.0	19.8	15.8	15.7
All public	20.2	19.8	19.7	19.3	9.6	9.2
Forest industry	0.0	0.0	0.0	0.0	0.0	0.0
Other private	0.2	0.3	0.3	0.5	6.2	6.5
Pacific SW	18.6	18.2	18.1	17.2	17.5	16.4
All public	9.9	9.9	9.9	9.1	9.6	9.1
Forest industry	2.2	2.4	2.7	2.7	2.8	2.8
Other private	6.5	5.9	5.5	5.4	5.1	4.5

Table 2.2 (continued)

	1952	1962	1970	1977	1987	2000*
Total Pacific Coast	83.8	81.8	81.8	79.1	72.2	70.2
All public	55.8	55.0	54.6	52.7	41.6	39.9
Forest industry	11.3	11.8	12.3	12.6	12.5	12.8
Other private	16.9	16.0	14.9	13.8	18.2	17.4
Total Rocky Mountains**	66.6	66.9	64.5	60.2	61.1	59.9
All public	49.5	50.0	47.7	43.5	44.1	43.1
Forest industry	2.2	2.2	2.2	2.1	3.0	3.0
Other private	14.9	14.7	14.6	14.6	14.0	13.8
Total	509.4	515.0	504.0	491.0	483.4	475.6
All Public	153.3	152.4	150.2	144.2	136.4	134.2
National Forest	94.7	96.8	94.6	88.7	85.2	83.8
Other public	58.6	55.7	55.6	55.5	51.2	50.4
Forest Industry	59.0	61.3	67.6	69.0	70.7	71.4
Other Private	297.1	301.3	286.2	277.8	276.3	270.0

*Forest Service projections. **Includes Great Plains states.
Source: U.S. Dept. of Agriculture (1982, 1990).

year-to-year status of forests with respect to the broader conditions of ecosystem health such as soil and watershed conditions, dependent species populations, and so on. At best, parts of the puzzle can be pieced together for particular forest areas from heterogeneous biological studies and from various efforts to quantify multiple-use management objectives.

The calculation and projection of growing stock inventory is, of course, a highly inexact science. In projecting timber growth, estimations must include the extent and nature of existing and replacement growing stock; the effects on rate of growth of various prospective levels of management (e.g., use of pesticides and fertilizers) among various classes of ownership; and even more highly uncertain variables such as the frequency of major droughts, fires, and infestations, as well as the long-term viability of timber plantations—where the natural defense mechanisms of forest ecosystems are compromised or lost. The negative effects of intensively managed plantations on forest soil productivity were noticed in the 1980s, for example, when there was a 25 percent decrease in the annual growth rates, and a sharp increase in mortality in third-generation southeastern pine plantations. Similar conditions have been widely observed in European plantations.[25]

Both global warming trends and the connection between air pollution and forest degradation also create considerable uncertainties in projecting future timber growth. The comprehensive 1990 U.S. Forest Service report on the U.S. timber situation, an assessment prepared each decade as a requirement of the Forest and Rangeland Renewable Resources Planning Act of 1974 (RPA), specifically identified the potential impacts of global climate change, air pollution, and acid rain as significant variables that may influence future timber supplies. Yet, although the report quantified possible effects of air pollution in generating alternative future timber scenarios, because of the "undetermined implications" of global climate change it did not incorporate any potential effects of climate change into quantified projections.[26]

For the preceding decade (1977–87), the following assessment of changes in growing stock volume was estimated by the U.S. Forest Service. The total growing stock inventory increased by about 4.3 percent, spread fairly evenly across the country in all regions except the Pacific and across all ownerships except federal: logging exceeded growth by nearly 8 percent in National Forests overall and by more than 9 percent in the Pacific forests overall. Indeed, between 1952 and 1987 the total softwood growing stock in Pacific forests declined by 23 percent. Throughout the 1980s, for example, California's industrial timberlands—including significant stands of old growth redwoods—were liquidated at an average rate of 175 percent of growth, and up to 300 percent of growth in some counties.[27]

On a national basis, the forest products industries obtain about 30 percent of their roundwood from industrial timberlands. Almost half the roundwood consumed comes from farmers and other nonindustrial private timberland owners and about 22 percent comes from public forests.[28] Within this broad framework, however, the consumption of roundwood by the two dominant sectors of the forest products industries—lumber and pulp—diverged sharply, especially as growth in pulp production outpaced lumber production. By the mid-1980s approximately 60 percent of domestic lumber production was based in the West, where the quality of timber was generally very high due to remaining old-growth forests, and about 30 percent was based in the South. With primary roundwood processing focused on lumber, western states produced only about 14 percent of all primary pulpwood, and western pulp producers obtained almost two-thirds of their wood from wood residues (thus, given the capital investment required in pulp and paper production, representing a very strong dependency on lumber production and an important indirect demand on timber production).

The South, by contrast, which is prominently characterized by secondary forests that resurged naturally after the area was extensively logged in the nineteenth and early twentieth centuries, and increasingly by tree plantations, accounted for nearly three-fourths of total primary pulpwood production, and southern pulp producers obtained only about a quarter of their wood from residues. By the late 1980s more than 68 percent of domestic woodpulp production had become concentrated in the South, with the West and North each accounting for only about 16 percent.[29] Central and northeastern states are characterized by a higher proportion of paper production and end-stage manufacture of finished products relative to pulp production.

The massive escalation and concentration of woodpulp production in the South has increasingly been followed by lumber producers. Although the situation in the West remains volatile and predicted levels of decline in permitted harvests cover a broad range, it is certain that recent (e.g., 1980s) levels of logging will be restricted on both private and public lands, declining by as much as a third or more by the year 2000.[30] Between 1978 and 1990 the seven largest U.S. lumber and plywood manufacturers increased production capacity by 121 percent in the South and decreased capacity by 35 percent in the Pacific Northwest. Georgia-Pacific led the way in 1982, returning its corporate headquarters from Oregon to the company's original home in Georgia and now owning more than 4 million acres of southern timberland. It has been followed by increasing investments in southern timberland holdings and production capacity by Weyerhaeuser, Boise-Cascade, and other integrated wood products companies.

Although the depletion of western forests and the associated restrictions on logging are certainly the most significant force behind the recent shifts in focus, the South has a number of advantages in addition to a climate conducive to fast-growing pines: a low-wage labor force, comparatively easy physical access to timber (relative to mountainous western conditions), the comparative fragmentation of state timber policies, and, not least, a high level of private timberland ownership. As one industry member observed, "The reception for generating income off of the land from timber, and keeping it as unregulated as possible is something that is much more accepted in the South than it is in the Northwest."[31]

Rather than taking decades to develop, however, the environmental opposition in the South has escalated rapidly as the threshing machine has come full circle. As Robert Hagler argued in 1994, "the U.S. South may well become the environmental battleground of the next decade."[32] At the same time new prospects for timber shortages have arisen due to the sheer magnitude of ever-increasing modern production levels. By 1988, when the U.S. Forest Service released a controversial study entitled *The South's Fourth Forest: Alternatives for the Future,* it was becoming apparent that the levels of harvest on nonindustrial private timberlands were exceeding growth and that fewer private timberland owners were managing their lands for timber production. This reflected, in part, a diversification in the reasons for private ownership of forests as well as what some have called a lack of understanding of the economics of private timber investment.[33] These and other factors lead to projections that the standing softwood inventory in the South could fall by as much as 40 percent by the year 2000.

The forest industries responded both with attempts to discredit the study and with aggressive plantation programs, adopting the increasingly widespread practice of draining low-lying swamps and coastal wetlands to plant pine trees, for which they have come under sharp fire. In 1991, for example, environmental groups filed a lawsuit under the Clean Water Act to prohibit Weyerhaeuser from draining a cypress swamp on the coastal plain of North Carolina. The South has even acquired its own version of the Northern Spotted Owl in the form of the Red-Cockaded Woodpecker, which nests in the cavities of southern pines. The Louisiana subspecies of black bear, and the Mexican Spotted Owl have also been, respectively, listed and proposed as endangered species and depend on southern forest habitats.[34]

In addition to continuing contention over western forests, including those east of the Cascades in the northern Rocky Mountain states, and to escalating demands on the private forests of the South, nonfederal public forests in various regions of the country are also increasingly affected by industrial appetites for wood. The state of Minnesota, for example, which leads the nation in nonfederal public timberland ownership, saw capital expenditures by wood-based industries increase threefold in the 1980s, ultimately leading to a major public outcry over harvest levels. Efforts to further stimulate aggressive management (timber production) in these nonfederal public forests were proposed in the 1990 Resources Planning Act, with the goal of off-loading some of the burden being placed on National Forests.[35]

Since at least the middle of this century conflicts over the management of public forests and, increasingly, private timberlands have been bitter, complex, and continuous, and they show no signs of abating as production continues to rise. Even with the crudest possible definitions of sustainability and with the greatest optimism about the future, we are quickly narrowing the gap between rate of harvest and rate of growth at the national level. The pace of production is significantly exceeding sustainable timber yield in some regions and rampantly continues the process of simplifying and diminishing forests on a vast scale.

Alternative Futures

The most common views of the prospects for ever-increasing timber production look to several trends. One assumes increasing levels of restriction on timber harvests from National Forests, which, combined with owl habitat conservation efforts that affect private lands as well, primarily influences western production levels. A second assumes more intensive management of both other public forests and all private timberlands for timber production and is substantially focused on southern timberlands although various northern and western regions are also affected to differing degrees. It is increasingly clear, however, that diverse forms of environmental opposition, along with a lack of investment in timber management by nonindustrial private timberland holders (many of whom increasingly include the preservation of wildland and species habitat among their interests), will not allow the South both to make up the loss of timber from western forests *and* meet forest industry production levels that continue to grow rapidly.

Various other developments may also play increasingly important roles, including prospects for new wilderness and parklands, as well as increased state and local restraints on harvesting private land. Reacting to the incremental effects of these continuously expanding environmental concerns on timber production, leading industry spokesmen have observed (drawing upon the language of the late Senator Everett Dirksen) that with "a billion here and a billion there . . . pretty soon we will be talking about a 'real' and substantial wood deficit." They argue that "an incremental, forest-by-forest and issue-by-issue 'adhocracy' threatens the national timber supply."[36]

The discussion at this point often tends to turn outward and to the long-term prospects—not presently being seriously challenged—for increasing imports of all wood products, but primarily wood chips, lumber, and pulp, thus increasingly extending the conspicuous environmental burdens of timber production to meet rising U.S. wood products production and consumption levels across international borders.

Indeed, the paper industry is already well into the process. In Honduras, for example, Stone Container, one of the top fifteen U.S. paper companies, was negotiating for rights to log, chip, and export more than 140 million cubic feet of wood annually from 2.5 million acres of softwood forest in the eastern Mosquito area. (A tentative agreement with the Honduran Department of Forestry, however, was provisionally suspended in 1992 due to intense environmental opposition.) In Chile, Scott Paper is involved in a joint venture that owns 80,000 acres of forest. In Canada, Weyerhaeuser possesses long-term licenses that allow harvesting of 9 million acres of forest in three Canadian provinces. In New Zealand, ITT Rayonier has acquired cutting rights to more than a quarter million acres of timberland. In Brazil, the mammoth market pulp mills built or under construction (with investments by U.S. and European paper producers and development banks) are increasingly poised to provide major infusions of woodpulp to U.S. paper manufacturers. U.S. lumber producers have a particular interest in the Siberian forests which—although they are the world's largest and most untouched standing softwood forests—

have inherently slow growth rates (perhaps a tenth that of the Pacific Northwest) due to the cold climate and soil conditions. For both political and ecological reasons, prospects for sustainable timber production in Siberia simply do not exist. Pittsburgh-based Snavely Forest Products has already established joint ventures in Latvia and Russia, and has begun marketing Siberian lumber to U.S. manufacturers. In Venezuela, Louisiana-Pacific has obtained access to a million acres of forest to make wood products for export to the United States[37] This is only a small sample of the U.S. forest products industry's foreign investments and does not include investments by foreign-based corporations in production geared toward U.S. markets.

As Lipke writes on the global impacts of proposed set-asides for the spotted owl conservation program, "For every economic loss in a preserving region, there are partially offsetting gains in other regions. For every environmental gain in a preserving region, there may be more than offsetting losses in other regions." The phrase "more than offsetting" indeed understates the realities of the worst forms of timber production in foreign forests where the governments are corrupt or weak, the domestic economies in disarray, and the potential for democratic social influence on the management of the resource effectively nonexistent. Con Schallau and Alberto Goetzl, economists for the American Forest Resource Alliance and the National Forest Products Association respectively (the latter now merged with the American Paper Institute), have built on this theme. They state, "This nation's material requirements should be evaluated from the standpoint of equity as well as from an environmental perspective," and "if the supply of timber in the United States decreases, we would have to increase our reliance on the Third World for raw materials." They reference Jim Bowyer, who cautions, "With regard to equity, it is important to realize that when we elect, by design or default, to have raw materials gathered and processed elsewhere, rather than in the United States, we are, in effect, exporting the associated environmental impacts."[38]

If the recent conversion of industry spokesmen such as Schallau and Goetzl to the cause of environmental justice is somewhat suspect (they are, of course, arguing for increased domestic timber production), the points they raise are nevertheless well taken and provide rather unexpected industry support for the argument suggested earlier: that unbalanced international trade in unfinished commodities that embody severe natural resource depletion, and in the case of pulp, high levels of industrial pollution, can be strongly opposed on both national and international environmental grounds.

Other industry spokespeople, however, ignore the premise that the environmental burdens of U.S. production and consumption of wood-based products are poised to flood into the southern hemisphere (where forests already are fast vanishing under the burdens of land-use conversion, resettlement programs, household energy demands, and the explosively rising demands for both lumber and paper in the newly industrializing countries) and elsewhere. They instead wax philosophical on the subject of Cornucopians and Malthusians, finding comfort in the postulation that the productivity gains created by sustainably minded Cornucopian timber technologists will erase all limitations. F. Keith Hall, for example, chief scientist of International Paper, the largest paper company in the world, emphasizes the productivity gains possible in U.S. forests. He cites the achievements of the plant biologists, the geneticists, and the pesticide and machinery manufacturers who made possible the productivity gains in American agriculture in the twentieth century, "without which Malthus might well have been right," and candidly observes:

> Until recently we in the forest industry have been largely hunter-gatherers of wood, acting at a considerably more primitive cultural level than even the 19th century U.S. farmer. We went where the wood was, took what we wanted, and moved on. This practice is now reaching its end as the last harvestable virgin forests in North America are used up. So we have become farmers.[39]

Hall describes the historical approach of the forest industries to the forest resource as only the first, and most primitive, of a sequence of steps that should now be taken toward optimal productivity. The second step is the practice of natural stand management using the techniques of selective harvesting and restocking. Step three moves to managed plantations with unimproved stock, steps four and beyond to managed plantations based on genetically improved stock. Most of the U.S. forests are managed in one of the first two steps, while step three is said to correspond to the level of agricultural technology in 1900. In Hall's words, step three "represents what we would regard as the minimum acceptable baseline against which future developments must be measured. . . . We can vastly increase U.S. forest productivity by applying known technology and we know how to develop the silvicultural systems to bring most of our U.S. forests to this third step."

Understanding something of the difference between forests and plantations, as well as understanding the environmental impacts of twentieth century American agricultural practice, may well cause one to shrink from this Cornucopian view, much as we today shudder in contemplating the rising export of unconstrained American demand to the forestlands of Brazil, Chile, Venezuela, Honduras, New Zealand, Indonesia, Malaysia, Vietnam, Siberia, Canada, and elsewhere. There are of course other alternatives, even if they are never or rarely discussed. However, to consider them we must begin to disentangle the paper industry from the other forest products industries. This is actually easier than one might think by now—at least at an analytic level—for several reasons.

Reconceptualizing Materials: From Wood-Based to Bio-Based

It is useful to reconsider the primary wood-processing industries in terms of a distinction between those that produce "solid-sawn" products (primarily lumber), and those that do not (wood-based panels such as particleboard and OSB, and pulp). The latter group is able to make products by reconstituting fiber fragments and particles and thus potentially to make use of a significantly broader array of fiber types and sources than the first, which requires logs. Not only wood residue from sawmills, but also recovered urban wood waste (e.g., from demolition and construction) represent usable fiber sources. Equally important, both wastepaper and nonwood plant fibers represent important alternative resources not only for pulp, but for other traditionally wood-based products. It is increasingly clear, as one analyst observes, that "wood is not the only fiber source available to the wood industry."[40]

In recent decades, with the growth in this second group of industries, a body of research has arisen that is characterized in part by a search for new terms to describe these "wood-based" industries and products in relation to this heterogeneous, nonforest-based, and even nonwood-based alternative fiber resource. Some researchers use terms such as *biobased composites* and *biobased materials industries.* Others favor the terms *lignocellulose composites* and *lignocellulosic industries.* (Any substance that contains both lignin and cellulose—all plant growth, or biomass—is a *lignocellulosic.*)[41] Although some of this research directly intersects with an interest in paper fiber sources, most has been focused either on building materials or on the creation of advanced biofiber composites for specialized uses.

However, even within this narrower framework that has yet to fully include the paper industry, the language of a paradigm shift is unmistakable. Many of the frustrations that invariably dog those working at the front lines of such a shift are also apparent. Some researchers speak of the predisposition, skepticism, and bias in both academic and industrial circles, and amongst the public at large toward the notion of building materials made from straw or wastepaper. Sometimes this can be traced to early failures at commercializing new products, but more often it speaks of the inertia of mature industrial systems built upon long-established patterns of materials use.[42]

One of the larger frameworks in which this wood-to-lignocellulose shift is gradually occurring can be loosely characterized under the general mantle of *materials policy* or *materials efficiency policy.* It is an idea reflected in the names both of new organizations such as the Alliance for a Sustainable Materials Economy, and the Materials for the Future Foundation, as well as old organizations like the 1952 President's Materials Policy Commission, and in many of the debates that have surrounded RCRA (e.g., the 1974 hearings on "The Need for a National Materials Policy").[43] In many modern uses, the concept of materials policy is strongly tied to issues of natural resource protection and/or solid waste reduction, and increasingly to pollution prevention and even industrial reorganization. More traditional uses of the concept are primarily limited to economic issues of national materials security and the ability to meet the (unchallenged) resource demands of producers.[44]

The modern environmentally and socially oriented interpretations of materials policy are the focus of this discussion. John Young, for example, succinctly argued the case for a materials efficiency policy (a "new materialism") in a 1994 *World Watch* article. Documenting the staggering quantities of materials that support our daily lives, he points out that they slip by largely unnoticed until they become a waste and must be disposed. Yet, as he notes:

> . . . most of the damage is done well before waste management even becomes an issue. Most of the environmental dangers lie in the immense complex of mines, smelters, petroleum refineries, chemical plants, logging operations, pulp mills, and other facilities that churn out raw materials and bulk commodities. . . . It is at the front end and middle stages of the economy . . . that most of the abuses occur.[45]

Equally focused on raw materials and commodities flows, William Cronon too observes the degree to which we have been ignorant of or indifferent to the chain of economic and ecological relationships that bring the basic materials of daily life to our cities: "If we wish to understand the ecological consequences of our own lives—if we wish to take political and moral responsibility for those consequences—we must reconstruct the linkages between the commodities of our economy and the resources of our ecosystem."[46]

Cronon, as an historian, describes in rich detail how these far-reaching commodity chains came to be constructed around and mediated by a particular urban center in his groundbreaking work *Nature's Metropolis: Chicago and the Great West* (1994). Young and others essentially argue that they must be restructured and that the total throughput of materials must be reduced. A materials throughput or materials flow perspective is somewhat similar to the systems orientation of a product life-cycle perspective in that it follows networks of material relationships from raw materials to postconsumer disposition. However, unlike life-cycle analysis, which takes a product or process as the starting point (and thus, typically, as a given to be optimized) an environmentally grounded materials perspective questions all uses of core materials in terms of the technologies of production, the products, and the overall levels of consumption.

In some respects, the notion of reducing the throughput of core materials, such as wood, in the economy echoes certain characteristics of the toxics-use reduction variant of the waste minimization and waste reduction debates in the 1980s. These debates led, in turn, to the concept of pollution prevention—now solidly established as the successor to the pollution control model (even if still in search of a universally accepted definition and mode of implementation). The essential difference between use reduction and waste minimization was one of focus: the first concentrated on the seminal points at which toxic chemicals were introduced into products and production (such as design phases), the second focused just slightly upstream of the point at which they were released into the environment as by-products of production or end-use. Whereas waste minimization advocates substantially focused on the fine-tuning of production efficiency, use reduction advocates and many of their pollution prevention successors became equally interested in materials substitution and in product and process redesign. Interdisciplinary studies in pollution prevention have often ranged far afield of particular production processes in the search for changes in behavior, institutions, and policy perspectives that go well beyond the technology focus of pollution control and waste minimization.[47]

In some cases these efforts, arising in many different industrial, use, and research contexts, have led back to a focus on the production of a core substance, as has happened in the case of chlorine and chlorine chemistry. Reducing the production and industrial use of chlorine is now viewed by many as an underlying measure of success in reducing the complex web of organochlorine pollution problems that reach into every sector of economic activity, from agriculture to aerospace to pulp bleaching.[48] In just such a way, reducing the throughput of wood in the economy—through virgin materials substitution and recovered materials use, greater production efficiency, and the questioning of certain products and patterns of consumption—has begun to be embraced by some sectors of the environmental community. In this view, *wood use reduction* represents an underlying measure of both social and environmental opportunity in areas that range from forest conservation and industrial pollution prevention, to municipal solid waste reduction. As the title of one recent publication proclaims, *Cut Waste Not Trees: How to Use Less Wood, Cut Pollution and Create Jobs.*[49]

This perspective is beginning to be reflected in new groups such as the Wood Reduction Clearinghouse, a nonprofit organization founded in 1995, and the Systems Group on Forests, a blue-ribbon panel organized to identify practical means to reduce the global demand for wood and wood products. A joint venture of the Rocky Mountain Institute and Global Futures, the Systems Group has sought to build on the pioneering work of Amory Lovins of RMI in energy efficiency, incorporating as a key element of its efforts the establishment of a wood-use reduction program within the Future 500 Roundtable.[50] In addition to improving the use efficiency of materials and end products, one of the central elements of wood use reduction—focused at the front and middle stages of the materials flow—is the diversification of the material resource base available to the forest products industries with both new and underutilized fiber sources and waste products. Part of this process is reconceptualizing the nature of the core material.

The real challenge, however, may lie not in fitting the alternatives into the existing industrial systems, but in finding flexible, democratic mechanisms for redirecting the industrial systems to fit reworked perceptions of the materials. This subject is explored further in following chapters, in particular with respect to the paper industry's use of both wastepaper and chlorine. However, the remainder of this chapter focuses on the use of nonwood fibers in paper production as one aspect of the wood-to-lignocellulose shift and wood use reduction potential.

Nonwood Fiber Resources for Paper

The formal debut of the tree-free paper movement in the United States can probably be dated to as recently as 1993, when Earth Island Institute, a prominent San Francisco-based environmental organization, published part of its quarterly journal on kenaf paper. This, and subsequent articles detailing efforts by other groups and the availability of tree-free papers from various small ecodistributors, gave national visibility to what Earth Island came to call a "tree-free revolution."[51] One can only surmise, however, that to the handful of North American experts on the use of nonwood fibers in paper, it must have looked like "*deja vu* all over again."

The use of nonwoods as a fiber source for making paper predates the use of wood by roughly nineteen centuries. As indicated earlier, China, one of the largest paper producers in the world, depends on nonwoods for about 60 percent of its total paper fiber supply and in 1993 had about 15.6 million tons of nonwood pulp capacity. India, which like China has scarce forestlands, was the number two producer of nonwood pulps with about 1.3 million tons of capacity. Worldwide, nonwood pulps (mostly from China) represent about 10 percent of virgin pulp production and 6 percent of the total fiber supply to paper production. The United States, however, has only a handful of small pulp mills that use nonwood fibers, all for specialty applications. Total nonwood pulp production capacity in the United States was estimated at less than 0.5 percent of total virgin pulp capacity in 1993.[52] The U.S. industry's disinterest in nonwoods is essentially replicated in all other industrialized countries, although it was not always this way. Even within the modern era of paper production in the United States (essentially defined by the use of wood) nonwood fibers played a significant role during the first half of the twentieth century.

Nonwood plant fibers suitable for papermaking are tremendously varied, but can be placed into four general categories of sources as depicted in table 2.3.[53] (1) *Agricultural residues* are harvest by-products of crops produced for other primary purposes such as cereal grains (e.g., wheat straw and barley straw), other food crops (e.g., corn stalks and sugarcane waste or bagasse), and crops grown for seed (e.g., rye grass straw and oilseed flax straw). (2) *Nonwood fiber crops,* or *industrial fiber crops* are grown specifically as a fiber source. Such crops include kenaf, true hemp (cannabis sativa), jute, crotalaria (sunn hemp), sisal, abaca (manilla hemp) and others. They also include crops such as cotton and fiber flax, widely grown for use in textiles, although papermaking applications of these fibers come primarily from the use of their industrial residues. (3) In addition to cultivated plants, various *wild plants* such as reeds and grasses have found application as paper fibers. As commercial applications develop, some may be cultivated. They include the papyrus plant (from which the word "paper" is derived), bamboo, esparto grass, and a host of others. (4) The final major source has historically been one of the most important and consists of a wide array of *industrial and post-consumer textile and cordage wastes* (e.g., pure cotton or linen textiles, garments, and manufacturing wastes, cotton linters, which are a by-product of cotton ginning, old rope, and many others).

Both textile wastes and straw, along with various other nonwood fiber sources, had a noteworthy presence in U.S. papermaking through about

Table 2.3
Estimated availability of nonwood fibers, 1995 (million bone dry metric tons)

		U.S.	World
Agricultural residue	*Cereal straws:*		
	wheat straw	76.0	600.0
	barley straw	7.0	195.0
	oat straw	5.0	55.0
	rye straw	0.4	40.0
	rice straw	3.0	360.0
	seed grass straw	1.1	3.0
	seed flax straw	0.5	2.0
	corn stalks	150.0	750.0
	cotton stalks	4.6	68.0
	sorghum stalks	28.0	252.0
	sugar cane bagasse	4.4	102.2
Other nonwood fiber residues	cotton linters	0.5	2.7
	cotton staple fiber	3.5	18.3
	other industrial, commercial and postconsumer waste tow, textiles and cordage	na	na
Nonwood fiber crops	*Stem fiber:* kenaf (hibiscus cannabis)*, jute, true hemp (cannabis sativa), sunn hemp (crotalaria), ramie, etc.	0.0	13.9
	Leaf fiber: sisal, henequen, maguey, abaca (manila hemp), etc.	0.0	0.6
Wild plants	reeds	0.0	30.0
	bamboo	0.0	30.0
	papyrus	0.0	5.0
	esparto grass	0.0	0.5
	sabai grass	0.0	0.2

*At least 4,500 acres of kenaf was being cultivated in the United States in 1994, with an estimated yield of 20–30 thousand tons.
Sources: Adapted from Joseph E. Atchison, "Nonwood Fiber," *Pulp and Paper* (July 1995); KP Products, Inc. (kenaf acreage).

the middle of the twentieth century, accounting for up to 13 percent of the total fiber furnish in the early decades of the century.[54] Joseph Atchison, a prominent authority on the use of nonwoods as a paper fiber in the United States, reports that straw pulping maintained a meaningful presence in the Midwest through World War II, when there were more than twenty-five mills still producing corrugating medium from straw. Factors that led to the closure of all these mills by the 1960s, however, included increased labor costs in collecting and preprocessing nonwood fibers and the rising dominance of the kraft pulping and chemical recovery system[55] (the following chapter discusses some complications in applying the modern kraft process to nonwoods).

Even while the use of straw as a paper fiber source was dying out, efforts were underway to introduce new industrial fiber crop and residue uses to U.S. paper production. In 1952, for example, because of a domestic shortage of newsprint, a U.S. Senate subcommittee on newsprint released a study entitled "Newsprint for Tomorrow," which considered various alternative fiber sources, including bagasse. Around the same time a subcommittee of the House of Representatives released its "Study of Newsprint Expansion," which also examined the possibilities for utilizing alternative raw materials in newsprint manufacture. These reports lead to an inconclusive series of studies by the U.S. Department of Commerce and the Bureau of Standards on using bagasse in newsprint production.[56]

The troubled history of bagasse newsprint commercialization has been well documented by Atchison, but by the 1960s studies nevertheless confirmed that certain processes were both technically and commercially feasible. Among the most successful processes were those developed by the Hawaiian Sugar Planters' Association in collaboration with the Crown Zellerbach Corporation in the early 1960s. They were substantially driven by the Sugar Planters' intensive program to find ways to utilize sugarcane by-products. The research eventually led to a blend of chemical and mechanical bagasse pulp from which a fully acceptable 100 percent bagasse newsprint sheet was produced in Crown Zellerbach tests. Previous failures of other projects, however, had led to an inability to obtain financing for pilot mills, and Crown Zellerbach ultimately never commercialized the process.

While prospects for bagasse paper died out in Hawaii and elsewhere in the United States, the use of bagasse in other countries accelerated. At present bagasse is being used successfully in nearly all grades of paper and paperboard and is, after various straw residues, the second most important nonwood fiber source for papermaking, with bagasse pulp production reaching almost 3 million tons in 1993. Ironically, more than three decades later, there is continuing interest in bringing alternative fiber pulping to Hawaii, which remains 100 percent dependent on virgin pulp and paper imports.[57]

Also beginning in the 1950s, the USDA undertook a program of research into the use of nonwood fibers suitable for papermaking. The results were presented in a series of technical reports collectively entitled "The Search for New Fiber Crops" and published between 1960 and 1971 in *Tappi* (the journal of the Technical Association of the Pulp and Paper Industry). After screening more than five hundred plants on the basis of yield, stalk density, cellulose content, fiber length, and, for the frontrunners, bench-scale pulping trials, the USDA concluded in 1960 that kenaf, a fast-growing annual plant native to Africa, showed the most potential among a number of promising nonwood fiber sources for pulp and paper production.[58]

Daniel Kugler of the USDA, a leading expert on kenaf fiber cultivation and use, describes longrunning efforts to commercialize kenaf-based paper production in the United States in terms of five eras. The first began in the 1940s, when kenaf was investigated as a replacement for hemp and other nonwood fibers in making cordage (the cultivation of hemp in the United States had been outlawed by the Marijuana Tax Act of 1937, and jute and abaca imports were cut off during World War II). The second era, also driven by public investment, began with the USDA's selection of kenaf as the most promising paper fiber and culminated in the publication of the 8 July 1977 edition of the *Peoria Journal Star* on kenaf paper. The third era was marked by private investment, as USDA efforts were redirected to energy-related research in the late 1970s. Collaborations between the American Newspaper Publishers Association and International Paper examined the technical and commercial feasibility of a kenaf-based newsprint plant, and in 1981 a private company, Kenaf International (KI), was formed to commercialize kenaf applications.

The fourth era began in 1986 as the USDA, with funding authorized in the Food Security Act of 1985, reentered the picture to collaborate with KI on the "Kenaf Demonstration Project." A successful commercial-scale demonstration of kenaf-based newsprint production was undertaken with kenaf grown in Texas, pulped in Ohio, made into newsprint in Quebec, and used to print the 13 July 1987 edition of *The Bakersfield Californian* (the newspaper company was one of the joint venture participants in KI) and a number of other newspapers across the country. Plans to build a new 600 tpd kenaf newsprint mill in Texas, however, were indefinitely shelved due in part to changes in the domestic newsprint market, which led to an inability to obtain project financing.

The fifth era began in 1990 and has seen the development of several commercial kenaf growing and fiber separation operations, with the material used in a variety of niche-product applications from poultry litter to potting soil. KP Products of New Mexico is now manufacturing about 100 tons per month of printing and writing grade kenaf paper from fiber grown in Mississippi and pulped through a time-sharing arrangement at a small integrated bleached kraft flax pulp mill operated by P. H. Glatfelter in North Carolina. A James River mill in Delaware has also recently begun experimenting with the production of kenaf paper.[59] These small-scale commercial applications are seen as a stepping stone to larger investments in kenaf-based products. Charles Taylor of KI sees the potential for kenaf production increasing from the roughly 5,000 acres now being cultivated in the United States, to a 1–5 million acre crop within the next fifteen to twenty years.[60] Yet, the fact remains that after decades of promising research and development efforts, kenaf pulp production has yet to be significantly commercialized, just as bagasse pulp production has failed to find an enduring commercial application in the United States.

Every candidate nonwood fiber seems to have its hardcore boosters, and hemp *(cannabis sativa)* advocates are among the most passionate. They argue that unlike kenaf, a tropical plant whose cultivation has thus far been mostly pursued in southern (cotton belt) areas of the United States, hemp can be grown throughout the country. (Kenaf authorities, however, note that several of the hundreds of varieties of kenaf can be grown in cooler climates.)[61] Hemp advocates also point to USDA interest that dates back to a 1916 article by a USDA scientist who argued the case for "The Manufacture of Paper from Hemp Hurds."[62] Hemp was also identified as showing "significant promise," ranked slightly below kenaf, by the USDA screening efforts in the 1950–60s.[63]

Although hemp fiber advocates point to the wide availability of nonpsychoactive (low tetrahydrocannabinol content) cultivars (such as have recently been permitted for experimental cultivation in the province of Ontario and various European countries), the case for hemp fiber has been complicated in the United States by its perceived connection to marijuana legalization interests. Several states, however, have recently been examining prospects for industrial hemp cultivation.[64]

Outside the United States, hemp is being used as a fiber for making paper and has been the subject of growing interest in Europe. In November 1995, Germany became the most recent European country to lift a ban on industrial hemp cultivation, responding to a strong prohemp lobby comprised of farmers associations, textile and printing industries, and, as one recent article notes, "veterans of the alternative and hippie scenes of past decades."[65] It is believed that more than 800,000 acres of hemp are being cultivated for industrial use in the northern hemisphere, and several hemp mill projects have been contemplated for the Netherlands and the U.K.[66] The most significant recent effort has been undertaken by the Dutch Ministry for Agriculture, which in 1990 initiated a four-year study on the use of hemp in pulp and paper production.[67]

The use of cereal straws and other agricultural residues is also considered to hold significant promise in the United States as well as in Canada and Europe. Wheat straw, historically important as a domestic fiber source, is the second-most plentiful agricultural residue (after corn stalks) in the United States, with an estimated annual availability of more than 75 million tons; however there are tens of millions of tons of other straws available as well. By one estimate, depending on growing practice and soil type, an average of more than 50 percent of harvested cereal straw is available as surplus, or roughly one metric tonne per hectare of cereal grain (0.4 ton/acre), with the remainder tilled into the soil to prevent erosion and preserve soil productivity.[68] Several efforts are currently underway to commercialize straw pulping as a supplemental fiber source for commodity paper applications in the United States. One is based on the use of rice straw residues in the Sacramento Valley of California, another on the use of rye grass straw in the Willamette Valley of Oregon.[69] Both are being propelled by air quality legislation that seeks to eliminate the current practice of burning the residues. The Oregon project builds on a regional interest in commercializing grass straw pulping that dates back to research conducted by Oregon State University beginning in the 1960s. As OSU researchers concluded in a paper published twenty-five years ago, well before the severe regional timber-related economic dislocations of recent years, "we recommend that the paper industry reconsider this valuable source of raw materials."[70]

The current discussion of nonwoods as a fiber source for papermaking in the United States and other developed countries has several prominent characteristics. The first is linked to the extreme heterogeneity of the different plants and the ways they might be pulped and used. It is difficult to make an accurate statement about any particular aspect of nonwood use in papermaking without qualifying it by fiber type, farming practice, region, pulp process, production scale, and intended application. The variation in properties and potential applications in papermaking appears distinctly greater among candidate nonwood fiber sources than among commonly used wood types. Further, the use of particular nonwood fiber sources in papermaking may sometimes be best conceived in combination with a variety of other industrial applications. This industrial integration is characteristic, for example, of the use of bagasse, cotton ginning by-products, and textile wastes in pulp production. It is also characteristic of certain applications of bast fiber plants such as kenaf and hemp, whose long outer (bast) fibers have different properties than the short inner (core) fibers. Depending on the requirements of the intended paper application and the flexibility of the pulping process, one may want to pulp the whole stalk or only the preseparated bast and/or core fibers. Any unused fibers can commonly be used in other commercial applications. Thus, in contemplating nonwoods pulping as a commercial enterprise, one must also sometimes contemplate the need to jumpstart a chain of new industrial linkages that may vary widely depending on the fiber.

A second characteristic of the discussion of nonwoods as a fiber source for paper is that although an increasingly vigorous debate is beginning to be waged around a range of potential environmental and economic benefits associated with nonwoods use in papermaking, it is strikingly characterized by the huge vacuum in supporting contemporary technical research. Advocates, propelled largely by a strong hypothesis, a broad vision, a relatively small collection of research papers, and the certainty that the status quo is unsustainable, face a chicken-and-egg problem in advancing the affirmative case. Partly, this relates to the negligible level of commercialization of nonwood pulping in the developed countries (of which most is for specialty rather than commodity applications), and from a corresponding shortage of empirical data. Most modern commercial experience with nonwood pulping for commodity applications has been based on older pulping technologies in less-developed countries, and thus cannot be easily translated to the more demanding pollution control regimes, the higher-wage labor markets, or the product quality standards of developed economies.

Additionally, as discussed in the preceding chapter, the paper industry has the lowest levels of R&D of any major manufacturing sector in the United States. Compounding this is the established industry's unmistakable coolness toward nonwoods as a fiber source, and its ability to control the research agenda. In the publicly financed (and industry defined) $45 million forest products technology research agenda (Agenda 2020) mentioned in the last chapter, for example, nonwoods merited only a one-sentence nod: "Alternative plant fiber will be increasingly viewed as a component of fiber supply but will remain small overall"—perhaps a self-fulfilling prophecy.[71] Although wastepaper processing technologies received a significant share of attention in the research agenda, nonwoods were ignored. Yet, presently available agricultural residues *alone* represent a significantly larger potential fiber resource than even wastepaper—indeed, one comparable in magnitude to current wood use by the industry.[72] With some 60–65 million acres of farmland presently idled each year, at a taxpayer cost estimated at up to $15 billion, the subsidy reducing potential of alternative cash fiber crops is equally compelling.[73]

USDA expenditures toward commercializing kenaf—the only major federal program supporting nonwood fiber use in papermaking—were estimated at about $1 million in 1994.[74] Yet, compared to the technologies for using wastepaper, already highly developed by the time the recycling movement of the 1980s emerged, the modern agricultural, fiber preparation, and chemical process technologies of nonwood use in papermaking are in need of substantial further R&D and environmental and economic characterization in the United States. Critical questions include who will advance this research and on what basis, who will define its parameters, and who will pay?

Perhaps the most interesting aspect of the discussion is the sheer diversity of arguments being put forth in support of nonwoods as a fiber source for papermaking. They freely cross traditional boundaries of social, economic, and environmental policy terrains. In the United States, environmental arguments linked to the wood use reduction potential of nonwoods have lately been most prominent. The hope is to achieve greater protection of established forests and to forestall the rising environmental impacts of intensive timber farming by shifting some timber demand to the huge, largely untapped nonwoods resource. In other industrialized countries with more pronounced wood fiber limitations, currently available nonwoods represent a way of reducing imports.

The potential benefits of tapping the nonwoods resource are most obvious in the case of using agricultural residue such as wheat straw, particularly when surplus residue is traditionally burned and creates an air pollution problem. Indeed, from both resource efficiency and waste management perspectives, some analogy can be drawn with wastepaper recycling. As with recycling there is a common-sense appeal in bringing value to a waste (one writer speaks of "having the cake and eating it too"). The web of potential advantages can become even more complex in particular contexts. Because of strengthening prohibitions against burning rice straw in California, for example, finding industrial applications for this material has become linked to the future of the industry in the state. This, in turn, has been linked to preserving the ecological benefits associated with the winter flooding program for the rice fields, which creates critical wetlands habitat for migratory waterfowl.[75]

The arguments for using nonwoods as a fiber source for paper become particularly complex when industrial fiber crops such as kenaf, hemp and others are being considered. They constitute not only an alternative to wood use and timber plantations, but also a potential substitute for heavily subsidized crops such as cotton and tobacco, which invokes further areas of consideration. As alternatives to timber plantations, annual fiber crops are said to be less capital intensive, less risky (because fire or infestations will wipe out only one season's investment rather than years of investment), and more flexible.[76] Several candidate crops are thought to be capable under certain growing conditions of supporting equivalent or greater amounts of pulp production per acre per year than many fast-growing wood species, possibly with lower levels of chemical, water and energy inputs required.[77] Yet, for the time being, such arguments remain sparsely documented and contested due to the lack of modern comparative research, and to the need for such analysis to be grounded in the specific soil types, climate, and agricultural practices unique to different regions. Indeed, an appropriate comparative framework that can meaningfully accomodate a broad array of issues has itself yet to be well-articulated.

As discussed in the next chapter, there is growing evidence that a powerful pollution prevention and energy efficiency argument might be advanced for nonwood pulping relative to woodpulping, based primarily on the generally lower lignin content of nonwood fibers in combination with other characteristic differences between major candidate nonwoods and reference wood types. Like the land-use and agricultural inputs arguments, however, the pulp production case, although compelling, cannot yet be well substantiated (for chemical pulping, although it is comparatively well established for mechanical pulping) due to immense gaps in experience and research. For purposes of comparison it must also be elaborated on a basis that is specific to fiber, pulp process, and intended pulp application. Nevertheless, in an era in which pulp producers are embarking on billions of dollars of long-term investment in woodpulp mills to further reduce pollution and improve energy efficiency, the wide availability of forest friendly fiber resources, which may be intrinsically easier to pulp and bleach, has captured rising levels of interest from the environmental community. It surely merits more than a wink and a nod from the established industry.

Not least, a whole host of arguments has built up around the potential for diversifying and renewing the agricultural economy. This has been a particularly prominent theme in Europe, which has seen significant surpluses of agricultural food and feed commodities, and where policy is actively focused on developing nonfood industrial applications for agricultural crops. As one analyst writes, "Farmers would be able to grow cash crops and become active members of the economy rather than passive receivers of subsidies."[78] The Association for the Advancement of Industrial Crops states, ". . . overconcentration and overproduction in a relatively small number of food and feed crops have created global problems. Clearly, diversification in agriculture is of high priority."[79]

Al Wong, President of Arbokem (based in Vancouver, British Columbia), a prominent technical authority on nonwood fibers and now operating a demonstration-scale wheat straw pulp mill in Alberta, speaks of a "new millennium" in agriculture for Canada. He writes that despite new trade agreements, "subsidies [e.g., to grain production and export] will be continued in different forms, through various exemptions, by many countries. Canada can not afford to fight the subsidy war against the United States and the European Union." Even with price supports and the increasing size of farms, it is difficult for many grain farmers to keep up economically. Yet, he also argues, "De-populating the countryside is not an acceptable way to rationalize agriculture," and "it is futile and a waste of money to provide more formal education and skill training without the concomitant creation of commensurable jobs in the rural economy." In his view, a critical element of the solution is the utilization of untapped agricultural residues, such as the almost 50 million tons of available Canadian wheat straw, to make value-added products such as pulp in rural areas. His model of low-impact "community pulp mills" based on agricultural residue is further explored in following chapters.[80]

Tom Rymsza, President of KP Products, the first domestic manufacturer of kenaf pulp and paper, argues that unlike long-term tree plantation investments, kenaf constitutes an immediate, viable cash crop alternative for many small farms:

> The farmer does not decide to plant trees or annual crops. The farmer has equipment (tractors, etc.) and mortgages, and an annual income is required to make payments to keep the farm operation going, and to pay salaries to the hired help who depend on farm operations for jobs and income. The farmer will either plant annual crops or cease conventional farming.

He also raises the significance of ownership issues associated with tree plantations versus annual crops from the perspective of the local farm community, observing that "the crop farmer and family are resident members of the rural community with a vested interest in water, air and quality of life issues." Although a farmer with no children willing or able to take over farm operations might sell the land or plant trees as a retirement strategy, more commonly:

> Another tree farm owner may be a timber company or a large paper user like Time, Inc. They buy or lease the land and plant trees. . . . As corporate, absentee owners who view the tree farm as an investment, their interaction with the community is vastly different than the farmer. They are financially able to purchase large tracts of land and wait a number of years to begin realizing income from that investment. This strategy impacts the local community.[81]

In Japan, the potential for using kenaf has been linked not only to the fiber requirements of modern paper production, but to the potential to revitalize the economies and cultural traditions of small villages, which for centuries practiced the art of fine handmade papermaking using nonwood fibers.[82] In Kentucky, prospects for renewing the historically important hemp industry have been associated with the strengthening antitobacco movement in the United States, and the potential for finding replacement crops in tobacco-growing states.[83] In the south of England, the nonprofit Bioregional Development Group has been examining the prospects for locally grown hemp and flax to provide a replacement for imported paper and textiles. They point out that "200 years ago almost every village in Europe grew its own flax and hemp" for cloth and seek to re-develop a "'bioregional' fiber industry—producing locally to supply local needs." They have developed a model for a small pulp mill using flax and hemp fibers as well as by-products of textile production and are seeking financing for a pilot plant. Goals include reducing imports, reducing transport (and therefore pollution and global warming), revitalizing the rural economy, and reducing the exploitation of the people and environment in developing-world supply regions.[84]

Two of the characteristics of nonwood fiber use about which there appears to be relatively widespread general agreement are the seasonality of annual crops and the transport costs of moving low bulk density agricultural fibers. From the established paper industry's point of view, both compare unfavorably to wood, which can be harvested year-round and stored indefinitely, and is much denser than agricultural fibers and thus comparatively less expensive to transport. Agricultural pulp mills continue to be contemplated (though not built) at scales of up to 500–600 tpd. However, there is increasing agreement that for reasons of transport efficiency and fiber supply security, the optimal scales for agricultural pulp production, even in developed countries, should be much smaller. Compared to the scales of modern kraft woodpulp production, they may be minuscule. Most recent proposals have examined pulp fiberlines operating at levels of about 10–300 tpd, and it has been suggested that the "economic radius" beyond which fiber transport becomes too expensive may fall in a range of roughly 100–200 km.[85] Generally, the transport issues associated with nonwoods, the corollary emphasis on smaller-scale pulping formats, and the issues of heterogeneity in fiber types, sources, and applications, are the basis of an increasingly strong regional theme that runs through the debate. It is reflected in the eclecticism of scenarios and arguments put forth in support of nonwoods as a papermaking fiber in different areas. Depending on the region and candidate fiber sources, extremely varied types of industrial, social, and environmental implications and opportunities are raised.

Overall, the dominance of the wood-based industry perspective and the associated research corpus has strongly tended to overwhelm the debate as it has emerged and to claim the benefit of the doubt. In 1995, for example, the Paper Task Force, a group convened by the Environmental Defense Fund in partnership with several major corporate paper users, released a study of nonwoods as part of its project on defining "environmentally preferable paper." Given the limited body of technical and commercial experience with nonwoods that is available, the report seemed surprisingly decisive in its conclusions that industrial fiber crops did not represent an environmental opportunity and in its generally discouraging assessment of nonwood fiber potential in papermaking. These findings and a clear forest industry-oriented analytical bias were among several aspects of the report that were sharply challenged by reviewers and outside groups. They constituted an important example of how a conventional wood-based perspective could undermine the nonwoods idea before it could even be argued.[86]

The paper had stated, for example, "This paper, like most studies, uses the large-scale wood-based chemical pulping operation as a point of reference," a perspective it had already attributed to the AFPA. The inappropriateness of confining the analysis to this perspective had been raised by two of the paper's original external reviewers, including USDA reviewer Daniel Kugler, who had stated, "use of the large scale wood-based chemical pulping operations as a point of reference, rules out virtually everything but wood-based operations." Due to significant controversy, including the withdrawal of three of the eight original reviewers from the process, the report was withheld from release pending revisions. The revised report, released seven months later, was much improved by a more thorough and thoughtful analysis that openly acknowledged both the limitations of available evidence and the depth of differences between various perspectives on the issue. This more careful, more qualified, and less prejudiced analysis led to the substantive reversal of core findings offered in the original paper. Indeed, key EDF staff appeared to have become quite enthusiastic about the prospects for nonwood fiber use in papermaking.[87]

When one surveys the breadth of issues and implications associated with the use of nonwoods in papermaking, it becomes clear that one's vantage point matters a great deal. If the question is one of fitting a heterogeneous and disbursed alternative fiber supply into a geographically concentrated, technologically rigid, vertically integrated, capital-intensive industry, one begins to better understand the modern history of nonwood paper commercialization efforts. The question becomes less why the commercialization of nonwood fiber pulping has yet to succeed, than why anyone in his or her right mind would still be trying. At best, nonwood fiber looks like a last-ditch supplementary resource to pursue, when and if the brakes are ever seriously applied to timber production. When one views the issue from a broad perspective of social and environmental opportunity, however, one sees nothing short of abject failure and gross irresponsibility reflected in the modern industrial status quo. One also sees an area of potential remarkable for its reach, its regional variability, and its human and ecological significance.

It seems fair to conclude that the established paper industry's lack of attention to nonwood fibers outrivals even its indifference to fully utilizing wastepaper as a fiber source until, in the latter case, public pressure forced the industry to reconsider. As Andrew Kaldor writes:

> . . . a commonly held view today among the pulp industry experts of developed countries is that the production of nonwood fibers is not viable or competitive in their economic environment. The same industries, on the other hand, are prepared to accept a heavy long-term reliance on wood fibers due to a perceived lack of alternatives.[88]

A theme constantly heard from the industry is the need for highly secure access to raw materials, which is not, after all, surprising in light of the industry's existing investment in production capacity. The streamlining and concentration of production, however, proceeded historically within the context of an increasingly integrated materials framework, leading to large-scale and inflexible production formats based on the use of wood. It seems ever more apparent that the endlessly resurging fears of timber shortage, the ever increasing degradation of natural forests, and the growth of sterile wood plantations arises primarily not from a lack of raw materials alternatives, but rather from a lack of structural flexibility to pursue them. There are many reasons for arguing that a more environmentally and socially desirable format for the pulp and paper industry would be decentralized and flexible, with an ability to accommodate and add value to regionally unique fiber sources.

There is little question, however, that the corporate efficiencies of scale and integration achieved (at the expense of both employment and environment) in modern wood-based pulping will not be easily duplicated in a more diverse, decentralized, and flexible industrial structure. The domestic industry, traditionally one of the world's lowest-cost producers, will certainly not choose such a path on its own, nor would it necessarily be economically rational to do so in the absence of coherent supporting environmental, economic, and technology policies that could help stabilize fundamental processes of raw materials transition and diversification in a mature industry. Yet, ironically, the industry's concentrated political power is such that it can and does effectively forestall prospects for such alternative policies by both direct opposition and by tenaciously clinging to their opposites. The list of contrary policies is so well known to the environmental community as to constitute a mantra: the continued national subsidy represented by below-cost timber sales from National Forests (most recently estimated at $1 billion in the period from 1992 to 1994), the failure to internalize real pollution and energy costs, the immense public subsidy to waste disposal, and so on.[89] As Young emphasizes, all such regressive domestic materials policies have increasingly harmful counterparts at the international level.

Although the paper industry—considered in terms both of its use of roundwood and of the additional demand it places on forest resources through its consumption of wood residues and integration with lumber production—is arguably the most significant single source of demand on domestic timberlands overall, its complex relationship to forest issues has often been relegated to a status of vague abstraction. At the same time there are clearly major opportunities to create a domestic paper industry that uses substantially less wood and significant advantages that might be gained from an industry based on regionally sustainable fiber resource use. Although the use of wastepaper has grown, it is effectively dwarfed by increased levels of production and consumption. The use of nonwood fibers has been mostly ignored, as has the potential and need to reverse the trends in consumption. The demand on domestic and foreign timberlands is ever climbing, yet our public policies remain grounded in a view of industry and the environment that has not changed appreciably in half a century.

Although there are many able analysts of forest policy and practice, those from the environmental camps have long tended to take a defensive preservationist approach. More lately, many have been drawn into the profound complexities of attempting to define and market certifiable sustainable forest management practices. Both approaches are clearly understandable, but they are insufficient. As long as the resource is treated in isolation from the causes and sources of the demand placed on it; as long as the demand fails to be cast not only in terms of rising consumption, but also in terms of rising production; and as long as the structural barriers of extreme industrial integration and concentration go unaddressed, the threshing machine will continue its travels both at home and abroad.

3

Production Processes and Pollution

In the early 1970s, before modern pollution control regulations took hold, a Ralph Nader study group investigated what it came to describe as the "unfettered dominance" of the state of Maine by a group of major paper companies. Along with the many forms of economic control the companies wielded over Maine communities and their manipulation of public institutions, the group documented the severe environmental burdens these manufacturers imposed on the state through both logging practices and production processes at mills. The book that resulted, *The Paper Plantation* (1974), provides in many respects a useful baseline gauge of what a relatively uncontrolled paper industry looks like. Referring to the impact on rivers, for example, they wrote, "All of the major rivers in Maine . . . have suffered environmental insults that would outrage even the most fainthearted of conservationists. . . . Every day Maine's sixteen mills pour out wastes that rob the rivers of oxygen, clog them with solid matter, and give them a vile smell."[1]

The industry released "staggering levels" of particulate matter and sulfur oxides to the air; some mills emitted up to 150 pounds of sulfur dioxide per ton of pulp produced. Noxious kraft mill odors caused "upset stomach, nausea, headache, eye and respiratory irritation, aggravation of illness conditions, interference with sleep, and reduction in appetite." Chemicals handling was often sloppy and resulted in frequent spills. A staff member of the Maine Department of Environmental Protection described one spill in which "150,000 gallons of white liquor [pulping chemicals] poured right out into the river. The pH shot up to eleven and the fish literally jumped out of the water." As bad as things were in the late 1960s and early 1970s, however, they had been even worse in earlier decades of modern chemical pulp production. In the summer of 1941, for example:

> Because of low flows and heavy pollution loads from sulfite pulp mills upstream on the Androscoggin River, the area became a stinking horror. Dissolved-oxygen levels dropped to zero and deadly hydrogen sulfide and other odorous gases emanated from the waterway as anaerobic conditions appeared. At times the searing fumes from the river tarnished metal and blackened the paint of houses as far as half a mile away.[2]

Pulp mills, along with oil refineries, chemicals and synthetic materials manufacturers, and primary metals processing industries such as iron, steel, and aluminum manufacturing, generally fall within a notoriously polluting class of manufacturing industries broadly identified as chemical processing industries—those that use chemical reactions and large quantities of chemicals as an integral part of their production processes. These industries, which supply the intermediate materials for much of the rest of the economy, are also typically characterized by a close association or integration with raw materials suppliers (timber, oil, ores) and very high levels of energy consumption, and they have been a primary target of the pollution control regulations of the past three decades.

Introduction to Pollution Control and the Paper Industry

The two most important federal laws to define the modern pollution control approach and its single-medium orientation have been the Clean Air Act (CAA)—first passed in 1963, but only really taking shape in the reauthorization and amendments of 1970—and the Clean Water Act of 1972 (CWA).[3] The Environmental Protection Agency, which administers their implementation, was created in 1970, as was the Occupational Safety and Health Administration, which performs similar functions for the workplace. Both the CAA and CWA approached the regulation of pollution by focusing on their respective media of concern: establishing criteria for air and water quality in the form of maximum concentration standards for particular, pervasive pollutants. In the case of air pollution sulfur oxides (SOx), nitrogen oxides (NOx), lead, particulates, carbon monoxide, and ozone—the so-called "criteria pollutants"—were targeted. In the case of water pollution, indicators such as biological oxygen demand, suspended solids, and others became the focus. Regulators then attempted to work backwards to the myriad industrial and other sources of those pollutants and to identify technology-based emissions and procedural standards that could be imposed to reduce releases to levels that allowed regional airsheds and water bodies to meet the overall ambient concentration standards.

By the mid- and late 1970s, the media of concern proliferated and became differentiated in response to gaps and failures in the original legislation and its implementation. These problems were addressed in such pieces of legislation as the Resource Conservation and Recovery Act of 1976 (RCRA), which regulated industrial and municipal solid waste disposal; and the Safe Drinking Water Act of 1974, which responded in part to the failure of the CWA to effectively control contamination of both surface water and groundwater; and in the rising attention paid to the phenomena of long-range air pollution transport, acid rain, and, more lately, global warming and ozone depletion.

Increasingly, these laws were also amended and supplemented in response to the substantial escalation of concern over toxic substances that, unlike most of the conventional pollutants first regulated, were largely invisible and posed risks of serious health and environmental damage in minute concentrations. The Clean Water Act was amended in 1977 to include a list of 126 toxic "priority pollutants." The Toxic Substances Control Act of 1976 (TSCA) was created to require testing and clearance of new chemicals before they were introduced into commercial use, as well to provide what has been alternately viewed as an "overarching framework" or a "gap-filler" for medium-specific toxic chemical regulation. CERCLA (usually known as *Superfund*) was passed in 1980 to address the cleanup of hazardous waste sites, and RCRA was amended in the mid-1980s to further elaborate the "cradle-to-grave" management of hazardous solid waste. Most recently, the Clean Air Act amendments of 1990 attempted to address the longstanding failure to control toxic air pollution by designating a list of 189 hazardous air pollutants to be regulated.

Reported expenditures on pollution control by the paper industry reflect this modern history of regulation in a characteristic pattern. During the 1970s, pollution control expenditures averaged 24 percent of capital spending. During the 1980s, however, as a function of both the expenditures of the 1970s and the deregulatory stance of the Reagan administration, expenditures on pollution control declined to an average of 8.1 percent of the industry's capital investments. With the major reauthorization of the Clean Air Act in 1990 and pending rules under the Clean Water Act, environmentally directed capital expenditures by the paper industry were expected to rise back into the 20 percent range during the 1990s.[4]

How these expenditures should be directed, however, and how they are even defined—for example, in relation to main business innovation spending—leaves much room for discussion. An "environmental compliance upgrade" by one name, may be a program for increasing production capacity, efficiency and profit by another. The complexities of defining the cost to industry of pollution control mandates have become even more pronounced as pollution prevention policy has begun to take hold. This newer approach tends, for example, to target investment decisions made well upstream of capital investments in "cleanup" technology. It is often substantially intertwined with basic operating efficiency and housekeeping improvements (such as better management of process chemicals inventories or better preventative maintenance) as well as with basic process improvements. These approaches may be as likely to yield a reduction in operating costs, or even in future liability costs, as to require a capital outlay.

There have been visible gains in air quality in most major metropolitan regions (even though many still remain out of compliance with air standards and are known as "nonattainment" areas) and similar improvements in the visible quality of many of the most polluted rivers and lakes. However, traditional regulatory approaches have been unable to meet the challenge of rising production levels and widening toxics concerns in many sectors, even with substantially stricter emissions standards for new facilities. Largely due to dramatically increased production levels, for example, the aggregate contribution to conventional air pollution by pulp mills in the United States increased substantially even after the CAA was implemented. For emissions of SOx, NOx, and carbon monoxide, the estimated contribution by pulp and paper mill production processes increased, respectively, by 73, 50, and 65 percent between 1970 and

1989, relative to overall decreases in each pollutant from industrial processes as a whole.[5]

Both water pollution and, generally, toxic pollution from pulp mills have proven even more difficult to control through traditional approaches. As with other manufacturing sectors, the first publications of the Toxics Release Inventory (TRI) data, beginning in 1989 (covering the 1987 reporting year), provided a shocking quantitative indication of the magnitude of chemical pollution associated with the paper industry on a national basis. The information from the TRI dovetailed with the already rising concerns focused on chlorine use in pulp bleaching and brought the industry into the limelight of national concern. The paper industry-focused set of "cluster rules," a largely unprecedented and aggressive regulatory foray into sector-focused, multimedia rule making, discussed later, was one of the outcomes.

Under the toxic chemical release reporting requirements of the Emergency Planning and Community Right-to-Know Act of 1986 (EPCRA, also known as Title III of the Superfund Amendments and Reauthorization Act, or SARA), manufacturing facilities that annually use more than 10,000 lb. or manufacture more than 25,000 lb. of more than three hundred toxic chemicals were required to report publicly their releases and transfers of these chemicals to each of various media including air, water, land, underground injection wells, publicly owned treatment works, and off-site hazardous waste TSD (treatment, storage, and disposal) facilities. The original list of chemicals (recently doubled) represented a highly arbitrary and incomplete selection of the several thousand or so chemicals that have been sufficiently well studied to have appeared on one or more of the major lists of toxic, hazardous, extremely hazardous, and/or carcinogenic chemicals devised under various federal and state regulatory and advisory programs. The high reporting thresholds have guaranteed that only the most major polluters will report and that only the most high-volume chemicals will be included. Undifferentiated total emissions are of only limited utility because they fail to reflect the potency and effects of each chemical. Nevertheless, even within these significant limitations, although the chemical industry has accounted for roughly half of all reported releases and transfers, the paper industry has ranked between second and fourth among major manufacturing sectors (vying with primary metals manufacturing for second place until sodium hydroxide and sodium sulfate—two of the most important chemicals used in pulp mills—were "delisted" in 1988).

Because of the large production scale and high-volume chemicals use of modern chemical woodpulping, many pulp mills ranked among the worst facilities in the nation by total TRI releases. In 1989, for example (the third reporting year and the first that was considered fairly stable) five integrated pulp mills were among the top fifty facilities in the nation by total TRI air emissions, and twelve were among the top fifty by surface water releases (most of the remaining top fifty were chemical manufacturers). Some pulp mills had estimated total TRI chemical releases of 5–10 million pounds per year. In some, air emissions alone ranged from the tens of thousands of pounds up to 2–3 million pounds each for chemicals such as acetone, ammonia, chlorine dioxide, chlorine, chloroform, hydrochloric acid, methanol, sulfuric acid, and others. The industry bore a disproportionate responsibility for releases and transfers of chemicals including chloroform (75 percent of all releases), polychlorinated biphenyls (PCBs, 18 percent), and friable asbestos (16 percent), all of which are human carcinogens. Altogether it reported releases of more than 25 million pounds of TRI-listed carcinogens, including formaldehyde, dichloromethane, tetrachloroethylene, chromium, and others.[6] Yet, the currently available TRI data, with its focus on a comparatively small number of high-volume chemicals, does not include many of the chemicals that have come to symbolize the concern over toxic chemical pollution from pulp production in the past decade: dioxins and furans, along with a host other highly toxic chlorinated organic compounds that result from the use of chlorine, chlorine dioxide, and other chlorine compounds in pulp bleaching.

Despite the billions of dollars invested in pollution control equipment and oversight by both the industry and government in the past decades, the TRI data, combined with investigations of organochlorine compounds, provided a nationally significant indication that dangerous by-products of pulp and paper production processes were still pouring into the environment on a large scale. Indeed, in certain cases the ongoing problems that result from pulp production, which plague many communities nearby or downstream from the one hundred and fifty or so chemical pulp mills in the country, have continued to sound remarkably similar to those of the 1960s and earlier decades described by the Nader study group. One news story describes the conflicts that sharply escalated in the late-1980s around pollution from Champion International Corporation's bleached kraft pulp and paper mill along the Pigeon River in Canton, North Carolina (just upstream from the Tennessee border):

> Now, as a result of eight decades of pollution, beyond Canton the once-pristine river has become a fetid sewer that Tennesseans say looks like Coca-Cola, smells like rotten eggs and is laced with potent carcinogenic dioxins. . . . In Hartford, a river hamlet of about 780 residents five miles west of the North Carolina border, townspeople blame the poisons in the stream for a rash of cancer deaths in recent decades. . . . In Newport, a town of 7,580 residents about 15 miles further downstream, environmental activists say fish taken from the river show high levels of dioxin, often have missing eyes and fins and are covered with unsightly sores. The town long ago stopped using the Pigeon for its municipal water.[7]

Using a classic jobs versus environment maneuver in the conflict, however, Champion (which has since undertaken a major multiyear modernization program) initially claimed it could address stricter pollution control standards proposed by the EPA only by laying off half its workforce, leading to a raging interstate feud between North Carolina and Tennessee.

These stories of economic hostage-holding, often accompanied by apparent regulatory paralysis and drawn-out litigation, have been witnessed in many communities and have long characterized the operations of the paper industry, in part because the large scale of most woodpulp mills often makes them the single-largest economic force in rural areas. In many respects, however, the conflicts are as much a failure of the environmental policy system as is the continuing pollution itself. Overall, the combination of serious local conflicts around particular mills and of the sectoral perspective provided by the TRI and the chlorine debates have led to a call for change and innovation in pulping technology on a far more radical scale than the incremental advances of past decades.

To provide the background necessary for examining some of these issues in more depth, the following sections begin with an overview of the major production processes employed in woodpulping and papermaking, and then introduce some of the issues and options that have begun to emerge out of new policy frameworks.

Basic Woodpulping Technologies

The two major approaches to woodpulping are broadly divided into mechanical and chemical processes, although various technologies combine the two. The primary constituents of wood are the cellulose and hemicellulose fibers that make up about two-thirds of wood structure on the average. The rest of the wood structure consists primarily of lignin, which provides structural reinforcement and binding of the cellulose fibers, as well as a small amount of various other nonfibrous oils, carbohydrates, and minerals.[8] Each pulping method is more or less adapted to the unique properties of different basic wood types, including hardwoods (such as eucalyptus, birch, poplar, beech, aspen, maple, linden, and alder); softwoods (coniferous trees) with low resin content (e.g., spruce, hemlock, and balsam fir); and softwoods with high resin content (most pines). And, of course, each process varies in terms of its immediate environmental impact, most importantly as a function of the amount and types of energy and chemicals required.

Mechanical Pulping

Mechanical processes decompose wood structure by grinding up whole logs or refining wood chips, and produce pulp that contains nearly all of the original components of wood. All mechanical processes thus deliver a very high yield of pulp (85–95 percent of input wood by dry weight). However, because the pulp retains lignin, which darkens when exposed to light, it is unsuitable for papers in which high whiteness or optical permanence are required. Pure mechanical pulping processes include *stone groundwood* (SGW) and *refiner mechanical pulping* (RMP). In stone groundwood pulping, the oldest mechanical process, logs are pressed against a revolving grindstone in the presence of water and, literally, ground to a pulp. The refiner method forces wood chips through rotating patterned steel plates. Because the cellulose fibers are fragmented and shortened in mechanical processes, they produce relatively weak pulps. They also require a large supply of electrical energy to drive the grinders. Refiner pulp is somewhat stronger than stone groundwood, but requires even more energy to produce.

Other processes that cluster near the mechanical end of the pulping spectrum include: *chemimechanical* (CMP), *thermomechanical* (TMP), and *chemithermomechanical* (CTMP and bleached CTMP or BCTMP) processes. Generally they use either chemicals (CMP), or heat and pressure (TMP), or both (CTMP) to soften the lignin before refining, thus obtaining improved fiber characteristics over pure refiner pulps. Virgin newsprint, which together with coated groundwood papers is the primary application for mechanical pulps, is largely based on TMP, with the addition of some chemical pulp to improve strength. Altogether, mechanical pulps accounted for about 10 percent of total domestic woodpulp production in 1993, a proportion that has been constant for several decades.

Chemical Pulping

Chemical processes are primarily characterized by a heated pressure vessel, called a digester, in which wood chips and chemicals are mixed and cooked to dissolve the lignin and to separate the fibers. The pulp is then washed and screened, leaving relatively purified and intact cellulose fibers and a small amount of residual lignin. The traditional chemical process alternatives are the *kraft* (also known as *sulfate*) *process,* the related *soda process,* and the *sulfite process*. Newer alternatives such as the *organosolve process,* and variations or hybrid versions of traditional processes have also begun to establish a commercial base. The yield from full chemical pulping ranges from about 40 to 60 percent of input wood by dry weight, although up to 65 percent in some modern sulfite processes.

In *kraft pulping,* sodium hydroxide (caustic soda) and sodium sulfide are mixed to create the "white liquor" in which the wood chips are cooked and digested. After cooking, the pulp is discharged to a blow tank, then washed and screened. The spent cooking liquor, or "black liquor," is transported to a recovery boiler where it is burned to yield recoverable thermal energy and an inorganic smelt of sodium carbonate and sodium sulfide. The steam energy from the boiler is used directly in mill processes and also in driving turbines to produce electrical energy that can be used by the mill or sold. The smelt from the boiler is discharged to a tank and dissolved to form a "green liquor." The green liquor is reacted with lime to convert the sodium carbonate to sodium hydroxide, which can then be used in a fresh cooking liquor. The recovery and reconstitution of inorganic kraft pulping chemicals, and the cogeneration of power from these combustive recovery processes, are integral to both the economic and environmental viability of the kraft process. However, the recovery system requires a substantial capital outlay and has been a central force driving kraft pulp mills to ever greater economies of scale. At present kraft mills are not considered economically competitive at levels much below 1,000 tpd.

The primary purchased chemicals for kraft pulping (not including bleaching chemicals) are sodium sulfate, sodium hydroxide and lime rock. The amount of make-up chemicals required relates directly to the loss of process chemicals to air, water, and solid waste streams. Thousands of other chemicals are used or formed as by-products and released in digester, recovery, effluent treatment, and maintenance operations. For example, various sulfur gases, including toxic hydrogen sulfide, methyl mercaptan, dimethyl sulfide, and dimethyl disulfide are released as byproducts of kraft pulping. They have long been a hallmark characteristic of kraft mills because even at very low concentrations (below one part per billion) they create a vile odor likened to rotting eggs. Indeed, although the kraft process was invented in Germany in the nineteenth century and the country is a major chemical pulp producer with a large softwood inventory, kraft mills have long been prohibited there primarily because of problems associated with emissions of odoriferous sulfur gases.[9] Similarly, air emissions of NOx, SOx, carbon dioxide and other conventional combustion by-products have traditionally been high, as with all fuel-burning, energy intensive technologies.

Kraft pulps, particularly long-fibered softwood pulps, have superior strength characteristics. However, the pulp is very dark in color (the brown paper bag of supermarket and lunchroom fame is primarily made from unbleached kraft pulp), with an unbleached brightness of about only 35 points on the ISO scale of 100. Conventional kraft pulps must thus be intensively bleached to attain the brightness desired in most papers used for printing and writing and in some grades of paperboard: generally above 65 points, with high-brightness papers around 80–85. Bleached kraft market pulp is typically bleached to a brightness of 88–90 points, which substantially exceeds the brightness commonly desired in printing and writing papers. It is in some respects a form of overkill that allows maximum flexibility in mixing kraft pulp with other pulps and additives in different papermaking applications. *Soda pulping*, based on the use of caustic soda or sodium carbonate without the addition of sulfur compounds, is a precursor of kraft pulping, with some modern elaborations discussed below.

Conventional *sulfite pulping* uses sulfurous acid and bisulfite ion (typically ammonium, calcium, or magnesium base) in the cooking liquor. It produces a characteristically light color pulp (60–70 points in brightness) with good strength properties although not equal to kraft pulp strength. However, sulfite pulping in the acidic regime does not disintegrate bark or wood pitch as effectively as kraft pulping and is not well suited for use with highly resinous softwoods such as pines. The conventional chemical recovery process for sulfite pulping is analogous, but somewhat different than the kraft recovery system. Unlike the kraft process, which is unprofitable unless chemical recovery is practiced, the one-time use of chemicals was historically affordable for sulfite mills. Until the 1960s, when pollution concerns escalated sharply, sulfite pulping liquor was dumped directly into rivers with little or no treatment. Some argue that the appalling environmental history of sulfite pulping has worked against the recognition of the benefits of modern sulfite technologies (described below) relative to kraft.[10]

Neutral sulfite processes (operating in the pH 9–11 range) have received attention in recent years due primarily to innovations in the use of quinone pulping catalysts such as *anthraquinone* (AQ), which accelerate the otherwise slow delignification rates of neutral pulping to commercially acceptable speeds. Applications include the neutral-sulfite-anthraquinone (NSAQ) process. A complex hybrid sulfite/organosolve process, the alkaline-sulfite-anthraquinone-methanol (ASAM) process is also being investigated in Germany. Not insignificantly, sulfite pulps are generally much brighter than kraft pulps, and high brightness can be more easily attained without the use of chlorine or chlorine compounds in bleaching. Alkaline-sulfite-AQ processes (operating in the pH 12+ range) can also produce 15–25 percent higher yields (and thus use less wood per ton of product) than kraft processes. They can process a wider variety of wood materials than traditional sulfite processes and produce pulp strengths competitive with kraft pulps.[11]

These newer sulfite or hybrid processes may offer one of many partial alternatives to the production of bleached kraft pulp from both hardwoods and softwoods. Although the high cost of the catalyst was initially viewed as a limiting factor, recent studies have investigated the synthesis of anthraquinone from recovered wood lignin itself. Anthraquinone has also been introduced into commercial use as a catalyst in kraft and soda pulping to improve delignification, obtain higher yields, and reduce the use of other pulping chemicals, albeit less substantially than in the alkaline sulfite regime.[12]

An alternative to both kraft and sulfite pulping processes, which has only recently entered commercial phases of implementation (although the basic idea has been around since the beginning of the century), is *solvent* (or *organosolve*) *pulping,* based on the use of organic solvents such as methanol and ethanol as pulping chemicals. The first commercial organosolve demonstration mill, using the trademark Alcell process developed by Repap Enterprises, began operating in 1990 in the province of New Brunswick. The Alcell process uses equal parts denatured ethanol and water to create an acidic cooking liquor and produces a higher-yield (hardwood) pulp than traditional kraft—and one that compares in strength and is easier to bleach.

Avoiding the elaborate chemical recovery systems required in inorganic chemical pulping, one of the chief characteristics of true solvent pulping is nevertheless the recovery of marketable wood chemicals from the spent cooking solution. These chemicals include furfural, a commodity chemical with numerous industrial applications (it once served as the raw material for nylon production until displaced by butadiene, which is currently produced from petroleum) and sodium- and sulfur-free lignin, currently being tested in various adhesive applications and as a water educer in concrete. As a basic polymer, lignin also has numerous other potential applications. Future marketable co-products may include vanillin, acetic acid, and ethyl acetate. Silvichemicals (wood-based chemicals)—including "naval stores" products such as rosin, turpentine, and fatty acid—have in fact long been produced primarily from chemical pulping by-products, with a value of more than $700 million by the early 1980s. A production format in which pulp mills produce chemical co-products is thus not unprecedented.

True solvent mills, however, do not have the benefit of the energy produced (primarily from lignin) by burning black liquor in a recovery boiler and will be heavy net energy purchasers. The value of the silvichemical co-products must therefore offset the increased energy purchase costs (in effect, the fuel value of lignin must be reassigned to its commodity chemical value). A key advantage of mills based on solvent processes, however, is that without the capital outlay required for the inorganic chemical recovery system they are expected to be competitive at less than a third the size of conventional kraft mills, or around 300 tpd.[13]

A dozen or so hardwood mills around the world are now using the *soda-AQ process,* which represents a sulfur-free alternative to kraft pulping. It is considered to produce weaker pulps than the kraft process from long-fiber softwood furnishes, but to be well suited to producing pulps from both hardwoods and nonwoods. However, the unbleached pulp is dark and is subject to the same intensive bleaching requirements as kraft pulps are when high brightness is desired.

Semichemical pulps, primarily used for corrugating medium, undergo a less-intense digestion phase (e.g, lower temperatures, shorter duration, less chemicals), which softens the lignin but dissolves only part of it, followed by a mechanical refining stage. Yields are typically in the range of 70–80 percent of the input wood by dry weight.

A special class of highly purified chemical pulps are derived from particularly intensive kraft or sulfite processes and are known as *dissolving pulps.* They are used to make rayon, cellophane, and various cellulose-based plastics and chemicals, and are thus used primarily outside of the paper industry.

Biopulping, although still in relatively early phases of development, has received increasing attention in recent years. It is not presently considered a replacement for other pulping or bleaching processes, but rather as an enhancement to reduce the use of energy and chemicals. Enzymes such as xylanase, which improve delignification in chemical processes, have already reached the commercial market. Various fungi have also been explored for application in pulping. A U.S. consortium of thirteen paper companies, the University of Wisconsin Biotechnology Center, and the U.S. Forest Products Laboratory began investigating biomechanical pulping in the late 1980s: using fungal pretreatments in combination with mechanical pulping.[14]

Early woodpulp production in the United States was based on nonresinous trees and mechanical methods. Production did not extend west to the predominately coniferous forests of the Pacific states until it was recognized that much of the resin in trees such as Sitka spruce, white fir, and hemlock is concentrated in the bark, and that the remainder of the tree could be effectively utilized for pulp.[15] Sulfite processes dominated chemical pulping in the early decades of this century; however the kraft method, when coupled with a workable chemical recovery system, opened up the production of chemical woodpulp to essentially any type of tree including the resinous pines prominent in southern U.S. forests.

In 1993 domestic woodpulp production capacity amounted to about 63 million tons. Almost 90 percent was based on chemical processes (see table 3.1). The kraft process (along with a small amount of soda pulping) accounted for almost 80 percent of all woodpulp and slightly less than 90 percent of all chemical woodpulp. Almost 60 percent of all kraft pulp, and all sulfite and dissolving pulps, were bleached. Sulfite and dissolving pulps each accounted for a little more than 2 percent of production, and semichemical pulp for slightly less than 7 percent. Hardwoods supplied about a third of the fiber furnish, and softwoods about two-thirds.[16]

Pulp Bleaching and the Chlorine Controversy

All wood pulps, depending on their intended end use, may be subject to a final bleaching process focused primarily on the remaining lignin. The substantial lignin content of high-yield mechanical and semichemical

Table 3.1
Woodpulp production capacity by process, 1993 (1,000 tons)

	Coniferous	Non-coniferous	Total
Mechanical	2,945	1155	3,101
Thermomechanical	3,405	179	3,584
Semichemical	0	4,171	4,171
Bleached sulfite	941	403	1,344
Unbleached kraft	20,205	1,064	21,269
Bleached kraft + soda	14,143	14,174	28,317
Dissolving	886	477	1,363
Total woodpulp	42,525	20,623	63,148

Source: FAO, *Pulp, Paper, and Paperboard Capacity Survey* (1994).

pulps can be brightened by altering the chromophores present in lignin. In this context common brightening chemicals include peroxides (such as hydrogen peroxide), and hydrosulfites (all totally chlorine free). In the case of full chemical pulps, bleaching refers to both the removal of residual lignin and the final brightening of the resulting pulp.

A traditional bleach plant sequence for chemical pulps begins by infusing the pulp with chlorine gas (also called elemental chlorine or molecular chlorine), which reacts efficiently with the non-fibrous materials (chiefly lignin), to form chlorinated compounds. This step is followed by an extractive alkaline treatment with dilute sodium hydroxide that dissolves the chlorinated compounds, which are then washed out. Final stages may include treatment with chlorine dioxide and hypochlorite plus additional caustic extraction stages. Chemicals that have more recently come into use in alternative chemical pulp bleaching approaches include oxygen, ozone, and peroxides.

Due to corrosion problems associated with the unwanted accumulation of chlorides in the mill process liquid system, bleach plant effluent is not normally recovered and recycled/incinerated like pulping liquor, and thus it has long been a focus of environmental concerns. For decades, these concerns have included the acute impacts of various chemical residuals and extreme pH values on aquatic life in receiving waters, high organic load (decreasing dissolved oxygen in water), and dark coloration. Despite improved methods for effluent treatment, a more ominous and far-reaching problem emerged in the 1980s. When organic matter, in this case primarily from the wood, combines with chlorine, a wide array of chlorinated organic compounds are formed. It is estimated that as many as a thousand different chlorine compounds may be present in conventional bleach plant effluent, of which roughly only a third have been identified.

Most of the handful of chemicals whose use has been banned or sharply restricted under various toxics laws in recent decades are synthetic chlorinated hydrocarbons including DDT, PCBs, toxaphene, and chlordane. They are generally stable artificial compounds—not easily broken down by natural biological processes—and can be extremely persistent in the environment. They are also fat soluble and accumulate in higher concentrations at higher levels in the food chain. Among the hundreds of organochlorines thus far identified in pulp mill effluent are such notorious compounds as carbon tetrachloride, chloroform, chlorophenols, and dioxins and furans. The latter are a family of compounds that have been indicted for reproductive, developmental, immune system and carcinogenic effects at extremely low doses, and subject to enormous attention. Generally, the organochlorine compounds from pulp mills that are of most concern are found among polychlorinated dibenzo-*p*-dioxins (PCDDs), polychlorinated dibenzofurans (PCDFs), and polychlorinated phenolic compounds (PCPCs). The most serious effects are presently thought to be associated with 2,3,7,8-tetra- to octachlorodibenzo dioxins and furans, along with some polychlorinated biphenyls (PCBs), and they have become a central focus of the "dioxin debates" (sometimes more appropriately called the "chlorotoxin" debates).

The problem is not limited to sludge and wastewater released by mills because dioxins and many other chlorinated hydrocarbons are also formed and emitted into the air when chlorine or chlorinated compounds are burned (at various process stages in mills or in municipal incinerators among other routes), and they have been measured in bleached paper products. Tests sponsored by the American Paper Institute in the late 1980s indicated that some disposable diapers had 2,3,7,8-TCDD concentrations of up to 11 parts per trillion (ppt, 10^{12}), paper towels up to 7 ppt, and paper plates, cups, and other food containers up to 10 ppt. The EPA has established 13 parts per quintillion (10^{15}) as the limit for 2,3,7,8-TCDD in water and has proposed limits of 10 ppt for combined 2,3,7,8-TCDD and TCDF in landspread pulp mill sludge.[17]

Various governments and some U.S. states have broadly sought to limit and even eliminate total organically bound chlorine (TOCl) or, more frequently, total adsorbable organic halogens (AOX) and have implemented or proposed color standards for effluent. Because the color is primarily influenced by lignin, and because digester lignin is burned (or recovered, in the case of solvent pulping) prior to bleaching, strict color standards may be functionally similar, in a roundabout way, to strict AOX or TOCl standards. They require either the bleach plant effluent to be recycled ("closing the bleach plant") or the prebleaching lignin content to be reduced very substantially. Both tend to imply reduced use of chlorine and/or chlorine dioxide in bleaching, although it is not guaranteed as it is with AOX or TOCl standards.

The domestic industry has tended to argue in favor of individual standards specific to the most toxic chlorine compounds (such as 2,3,7,8-TCDD) and against more comprehensive standards such as those based on AOX and TOCl limits. AOX and TOCl standards broadly encompass hundreds of these chlorine compounds, including the most toxic, the least toxic or even benign, and the many compounds for which comprehensive environmental and toxicological data is not available. To some extent, the former approach to standard setting, based on exhaustive chemical-by-chemical assessments, can be more closely associated with pollution control and the latter—more comprehensive and preemptive approach—with pollution prevention. The cluster rules for pulp mills, proposed in 1993 by the U.S. EPA, include a restrictive AOX standard for total mill effluent as well as individual standards for chemicals such as chloroform, pentachlorophenol, and TCDD in bleach plant effluent.[18]

The larger problem of organochlorine pollution is not limited to the paper industry, but encompasses one of the most far-reaching toxics problems of modern industrial society. It first arose in connection with pesticides such as DDT and others mentioned above. With the explosive growth in chemical production in the second half of the century, the problem has since extended to include the high levels of organochlorine pollution that arise from the manufacture of a vast array of chlorinated organic chemicals and plastic resins (most significantly, polyvinyl chloride or PVC); the incineration of plastics and other chlorine-containing materials in municipal solid waste incinerators; the notorious problems of chlorofluorocarbons as ozone depleters; the widespread use of chlorinated solvents such as methylene chloride in a huge variety of applications; and even the use of chlorine as a water disinfectant, where it is associated with trihalomethane formation.

These diverse concerns ultimately led to calls to phase out the use of chlorine and chlorine compounds in most industrial applications. Greenpeace has been the leading voice of this perspective, but the idea has also been endorsed by groups that include the American Public Health Association and the International Joint Commission (U.S./Canada) on the Great Lakes. The IJC, in a controversial 1992 recommendation (to which it has held despite enormous industry opposition) recommended a phaseout of chlorine as an industrial feedstock. Likewise, in 1993 the

APHA stated, "The only feasible and prudent approach to eliminating the release and discharge of chlorinated organic chemicals and consequent exposure is to avoid the use of chlorine and its compounds in manufacturing processes."[19] The paper industry has long been the largest consumer of chlorine aside from the chemical industry itself, accounting for nearly 20 percent of domestic chlorine consumption prior to the rise of the dioxin issue in the 1980s. It remains the second largest consumer, although its share had declined to about 9 percent by 1994.

The controversy over the use of chlorine in pulp bleaching came to focus on kraft pulping for several reasons that may be evident by now. For one, it is by far the dominant process and is currently expected to remain so; as such, kraft pulp has come to be known as "the pulp of reference" for the paper industry. It is also a dark pulp, requiring up to 100–150 pounds of chlorine per ton of pulp bleached in conventional bleaching. The use of chlorine also emerged as a particularly intractable problem for kraft mills because they are generally the largest, most expensive, and most efficient mills and are correspondingly less well adapted to accommodate process changes. These problems were compounded by the absence of a single direct substitute for chlorine use in kraft pulp bleaching, combined with a general weakness in R&D within the domestic industry. At the same time, as discussed further in the next section, the chemical industry has itself played a role in seeking to maintain one of its major chlorine markets, an effort assisted by a set of issues associated with the close dependencies between sodium hydroxide and chlorine production and consumption. These difficulties have been abetted by the traditional and entrenched risk-based pollution control framework. Indeed, the debates over "good science" and "acceptable risk" with respect to specialized analyses for potentially hundreds of substances, most of which are measured at the outermost limits of modern analytical chemistry techniques, have reached their highest form of art in the chlorotoxin polemics.

Although dioxin had been identified in the 1950s and used as an herbicide in subsequent decades, its notoriety as a toxin did not develop until the wide publicity attending the Agent Orange, Love Canal, Times Beach, and Seveso, Italy catastrophes in the 1970s and 1980s made it almost a household word. The EPA had begun to suspect a link between pulp mills and dioxin as early as 1980, which was substantiated when studies of fish caught downstream from several pulp mills in Wisconsin and subsequent effluent sampling revealed high levels of dioxin contamination in 1983. The rapid escalation of public concern over dioxin contamination from chlorine use in pulp bleaching, however, did not occur until the information became public several years later. In 1987 a Greenpeace report, "No Margin of Safety," researched and written by veterans of the struggle in the 1970s to eliminate the use of dioxin-containing herbicides by the U.S. Forest Service, documented the pulp mill/dioxin connection based on the EPA dioxin studies, which they obtained under the Freedom of Information Act. Subsequently, someone inside the American Paper Institute sent a collection of documents to Greenpeace that revealed a high degree of collaboration between the EPA and the paper industry to "suppress, modify, or delay the results of the joint EPA/industry [dioxin] study or the manner in which they are publicly presented."[20]

While the EPA and the U.S. industry (including both pulp and paper producers as well as chlorine manufacturers) have been engaged in elaborate studies and risk assessments designed to establish permissible levels for dioxin and other organochlorine emissions, a wide variety of alternatives that provide for the elimination of all chlorine compounds in pulp production have been implemented in European countries that moved aggressively to address the problem from a precautionary perspective. Lately, the debate in the United States has been cast in terms of whether elemental chlorine free (ECF) bleaching, based on the use of chlorine dioxide, or totally chlorine free (TCF) bleaching, where neither elemental chlorine nor any chlorine compounds are used, constitutes the appropriate goal. Overall, the recent levels of research and innovation driven by organochlorine concerns amount to a flood relative to the traditionally glacial pace of change in the industry.

The problem of organochlorine pollution from chemical pulp bleaching has come to be approached in several general ways, many of which may provide an array of other advantages, such as reduced energy consumption, reduced operating costs, and higher yields. The three major focal points are: (1) changing the pulping process (primarily to achieve a higher level of delignification), (2) changing the bleaching process, and

(3) improving the pollution control technology. The range of opportunities available, however, strongly depends on how the goal is defined. Reducing or eliminating the use of chlorine and chlorine compounds in *pulp production* opens up a far broader set of opportunities than when the goal is framed as one of reducing or eliminating the use of chlorine and chlorine compounds in *bleached kraft woodpulp production.*

Nonkraft chemical pulping processes, such as the solvent and the alkaline sulfite-AQ processes described above, all generally achieve improved brightness and/or more reactive (i.e., more easily extractable) residual lignin than when the same fiber is pulped in a kraft process. All such pulps can already be bleached in ECF processes and appear well adapted for TCF bleaching. Tests conducted on NSAQ pulped aspen, for example, bleached in a two-stage process using only oxygen and peroxide, have consistently produced a brightness of 85–88+ points.[21] These modern nonkraft processes also all produce pulps with kraft-competitive strength factors, usually at higher yields than kraft pulping, although to date only the ASAM process has been well demonstrated with softwoods. All have yet to be commercialized in the United States. The significant and rising dependence on pines—least suited to the use of nonkraft processes—is one of many barriers. (The prospective merits of nonwood fiber in a nonkraft TCF pulping regime are discussed later.)

Various modifications to the traditional kraft process, including longer cooking times and multiple infusions of fresh white liquor to the digester, are commercially available, although they may involve costly modifications to facilities. The combination of the two has become generally known as *extended cooking.* The primary objective of these digester modifications is to reduce the kappa number (a variable related to the lignin content of the pulp) below the traditional 30, to the range of 10–20 needed to pursue effective TCF or ECF bleaching. The capital costs associated with replacing or retrofitting digesters for lower-kappa cooking (in a 1,000 tpd kraft mill) are commonly estimated in the range of \$15–45 million. However, various modifications in operating parameters may be sufficient and available for newer digesters at comparatively minimal capital cost.[22]

The use of *oxygen delignification systems* as a prebleaching stage had already become widely adopted in Europe by the mid-1980s. U.S. pulp manufacturers have been comparatively slow to follow suit, although between 1990 and 1994 more than twenty new oxygen delignification systems were installed in U.S. mills.[23] Functionally, they are an extension of the digester process, focused on lowering the kappa number without damaging the fibers. Capital costs of new systems in a representative kraft mill are commonly estimated in a range of about $20–30 million. Various other mill changes—such as improvements in woodchip preparation and thickness control, and expansion of the recovery boiler capacity to accommodate increased organic material from extended pulping or oxygen delignification—may also be necessary or advantageous. Depending on the mill, capital costs may vary significantly, from a few million to tens of millions of dollars.

Alternative bleaching approaches range from those involving minor modifications of existing processes, to more radical substitutions. In the case of kraft woodpulp mills, TCF bleaching essentially requires that at least one of the above modifications (extended cooking or oxygen delignification) has been made, so that the lignin content of the pulp entering the bleach plant is substantially reduced below conventional levels. The most widely adopted change in bleach plants has been the partial or total *substitution of chlorine with chlorine dioxide* in the first bleach stage, which has been the primary factor in recent declines in chlorine use by the domestic paper industry. Major capital costs directly associated with chlorine dioxide substitution are primarily related to the need to expand chlorine dioxide production capacity (it is impractical to safely store and transport chlorine dioxide, thus it is customarily produced on site from sodium chlorate). Other costs may arise if modifications in process control systems are required. In most cases, upstream improvements in delignification will allow a reduction in chlorine dioxide use at the end stages of bleaching and free up chlorine dioxide to be directed to first-stage substitution. If substantial additional capacity is required, however, costs can be very high, ranging up to $15–20 million.

The costs of digester modification, expanded recovery boiler capacity, oxygen delignification systems, and chlorine dioxide capacity are the major capital expenditures that may be required to move a traditional bleached kraft mill in the direction of ECF pulping. Yet, depending on the age and condition of the mill, such expenditures may range dramatically, from a few million to $50–60 million or even more. However, many of these changes may also reduce operating costs. Savings come primarily from reduced chemical costs, but can also come from reduced energy consumption and in some cases from higher yields. Some have suggested cost savings may be substantial, exceeding $20/ton in an ECF process with oxygen delignification and up to $25/ton in a TCF regime using both oxygen delignification and ozone, relative to a benchmark mid-1980s bleached kraft mill with partial chlorine dioxide substitution.[24]

The comparative environmental advantage of chlorine dioxide substitution, which reduces but does not eliminate organochlorine by-product formation, is presently at the heart of the current debates in the United States. The ECF/TCF debate is complicated by one faction arguing that TCF bleaching will not be necessary to achieve *totally effluent free* (TEF) mills—a major goal in both North America and Europe. Effluent from both the digester and from oxygen delignification systems is customarily routed to the recovery boiler; however chlorides formed from chlorine or chlorine dioxide use prevent recycling of bleach plant effluent because they accumulate and can lead to severe corrosion of process equipment. Common wisdom is that TCF pulping must be practiced before "bleach plant closure" (essential to achieving TEF mills) will become possible. Yet, some, defending chlorine dioxide use, suggest chlorides can be isolated from bleach plant effluent and destroyed or recycled for chlorine dioxide generation.[25] Others view efforts to close the bleach plant in the presence of chlorides as more likely simply to shift organochlorine problems from one medium to another (e.g., to air, if bleach plant condensate is incinerated).

Increased chlorine dioxide substitution for chlorine (whether partial or total, i.e., ECF) is typically accompanied by other changes in traditional bleaching, including increased use of hydrogen peroxide and/or oxygen as reinforcements to the extraction stage. In some cases, modifications in the way chlorine is injected in the bleach plant have been pursued. Largely pioneered in the United States by Westvaco, this approach does not substantially reduce chlorine use, but reduces the formation of dioxins by injecting chlorine in multiple charges under higher pH conditions.[26]

The *use of ozone* has also recently entered full-scale commercial application in the Unites States—after several decades of fluctuating levels of interest and research—with the start-up of Union Camp's trademark C-Free ozone bleaching process in 1992. This process uses oxygen delignification, followed by ozone, extraction, and chlorine dioxide stages. Filtrate from all but the last stage is recycled, and a TCF version of the process was being tested.[27] The use of ozone, involving capital costs of up to $30 million, appears likely, as with oxygen delignification, to be a significant factor in TCF kraft bleaching regimes.

Bleach plant stages are typically designated by letters as follow:

C chlorine, Cl_2
D chlorine dioxide, ClO_2
E alkaline extraction—sodium hydroxide, NaOH
E_{OP} or (EOP) extraction reinforced with oxygen and peroxide
H hypochlorite, NaOCl or $Ca(OCl)_2$
L hypochlorous acid, HClO
N nitrogen dioxide, NO_2
O oxygen, O_2
P peroxide, H_2O_2 or Na_2O_2; hydrosulfite, $Na_2S_2O_4$
Q or T or X, chelants such as EDTA and DTPA
S sodium bisulfite, $NaHSO_3$
Z ozone, O_3

Conventional bleaching sequences include the classical five-stage CEDED process, along with CEHD, C(EH)DED and other variants. Alternatives that partially or totally replace chlorine with chlorine dioxide, and often add oxygen and/or peroxide reinforcement to the extraction stage include (CD)(EO)DED, (CD)(EO)(DN)D, and the ECF sequences D(EOP)D, and O-Z(EOP)D (the Union Camp ozone process). Totally chlorine-free sequences include O-(QZ)Q(PO) and O-Q(EOP)PPS (used at Louisiana-Pacific's Samoa, California mill). TCF bleaching for nonkraft and/or nonwood chemical pulps may be as simple as QP.

The pulp bleaching/dioxin issue, which broke in both North America and Europe at about the same time, has been the subject of close attention and extensive research by the industry, regulators, and environmental groups for approximately a decade now. Nevertheless, the regulatory and industry approaches to the problem have diverged noticeably between the United States, Europe, and Canada. In January of 1992, for example, British Columbia became the first government in the world to set a deadline for the elimination of all organochlorine compounds from pulp mill effluent, establishing a "zero AOX discharge" standard to be met by the year 2002. In the same year, B.C.-based Howe Sound Pulp and Paper (jointly owned by Canfor, a Vancouver-based forest products company, and Oji Paper, a major Japanese producer) became the first mill in North America to produce TCF bleached softwood kraft pulp. The province of Ontario followed suit in 1993, adopting the same zero AOX standard and timeline.[28]

In 1991 the Swedish company Södra Cell, one of the largest market pulp producers in Europe, announced its belief that "chlorine chemical bleaching has no future within the pulp and paper business," stating "this development is an opportunity rather than a threat." By the same year there were ten kraft market pulp mills with TCF capability operating in Europe, with more expected to begin operations shortly. In Sweden (the largest chemical pulp producer in Europe) all use of elemental chlorine for bleaching had ceased by 1993, the elimination of all chlorine compounds from pulp bleaching is expected to be achieved before 2000, and it is likely that the first closed bleach plants will be achieved at kraft mills within the country.[29]

In the United States, however, the first TCF bleached kraft pulp mill, Louisiana-Pacific's Samoa mill on the Northern California coastline, just began TCF operations in 1993. Moreover, Louisiana-Pacific's decision to move to TCF was the outcome of a series of regulatory sanctions culminating in its loss of a major lawsuit brought under the Clean Water Act by the Surfrider Foundation (the mill was dumping tens of millions of gallons per day of heavily polluted effluent into the ocean—at one of the better surfing spots in the state). The judgement carried a $5.8 million fine, at the time the second largest ever levied under the Clean Water Act, although Louisiana-Pacific was subsequently fined $6.1 million for violations associated with its Ketchikan, Alaska pulp mill. The 1991 settlement resulted in a consent decree requiring substantial improvement in the Samoa mill. The mill is not simply required to meet effluent standards, but also is required to demonstrate, through on-going bioassays, that several ocean species (including abalone, sea urchins, and giant kelp) can survive and reproduce in the discharged waste—an effective reversal of the conventional burden of proof.[30] As of 1996, however, LP had yet to be joined in TCF pulping by any other kraft mill in the country.

Reflecting the underlying connections between chlorine problems, the dominance of conventional kraft pulping, and the strong dependence on softwoods, the American Forest and Paper Association has attacked proposals for an AOX discharge ban in the United States, arguing, "No kraft mills have succeeded in making high-brightness pulp from southern softwoods without using chlorine."[31] In a particularly telling and widely publicized development in the spring of 1992, Georgia-Pacific, the largest domestic producer of bleached kraft market pulp, stolidly announced in a letter to its customers that it would not market TCF pulp, having decided to invest in chlorine dioxide substitution instead, and that customers who wanted TCF pulps would have to seek other supply sources.[32]

Overall, the industrial research agenda and rhetoric in the United States has seesawed wildly. Sometimes it offers the "grasshopper defense" (grasshoppers are reported to produce a chlorinated organic compound as an ant repellent): generally a line of investigation and argument positing that because some organochlorine compounds occur naturally, we shouldn't overreact to synthetic organochlorines with broad-brush strategies like AOX limits. Sometimes it makes more creditable efforts in technology development.[33] Hot-button phrases such as "minimal environmental impact," "no-effect pulping," "dioxin abatement," "environmentalist hysteria," "vilification of chlorine," and "acceptable risk" are pervasive. The need for "flexibility," the dictates of a "free market society such as ours," and the historical and cultural aversion of Americans to "march in lock-step" have all been advanced in defense of the use of chlorine and its compounds.[34]

It would not be inaccurate to conclude that, except for a handful of producers, the U.S. paper industry has been dragged kicking and screaming to the threshold of ECF pulping. It appears even more unlikely to go gently into the era of TCF pulping. In an astonishing keynote address at the 1995 International Non-Chlorine Bleaching Conference in Amelia Island, Florida, Helge Eklund, the President of Södra Cell, lashed out at American producers for "fighting like crazy to preserve obsolete bleaching technology." Referring to the subtitle of the conference, "ECF vs TCF—A Time to Assess and a Time to Act," he opened his remarks by stating, "Let me tell you one thing. Assessment has taken too long. Act now. Many of you have been sitting there doing nothing until legislation forced you to move." In a blistering indictment that could hardly have been surpassed by the most ardent Greenpeace campaigner, he asked:

> What are you doing in North America? . . . What are you waiting for, you technicians and experts who have not yet gone over to TCF? You must dare to suggest investments that do not immediately give you a great return . . . You must think long term . . . You are personally responsible for the pollution from your pulp mills, at least morally.[35]

Morally? It is not a word often used in the context of the paper industry. Yet Eklund's questions bear examination. It is worth stepping back to consider the different regulatory and industry settings that have yielded such different national responses to a globally significant problem.

Different Contexts for Innovation

Nicholas Ashford and others from the Massachusetts Institute of Technology's Center for Policy Alternatives have been investigating the relationship between regulation and technological innovation since the 1970s, and argue for the use of regulation to change the market for innovation. They contrast *technological innovation,* "the first commercially successful application of a new technical idea," with *technological diffusion,* "the subsequent widespread adoption of an innovation by those who did not develop it." Technological innovation is further differentiated in terms of process- and product-focused innovation, and in terms of major or "radical" innovation and "incremental" innovation. *Invention* is the discovery of a new technical idea.[36]

The various responses to the problem of chlorine use in pulp bleaching can be usefully considered in light of these definitions. In the United States, where the favored approach has been chlorine dioxide substitution, the response has been primarily one of incremental innovation and diffusion. The radical innovations associated in particular with oxygen delignification systems, ozone, and solvent pulping have been overwhelmingly concentrated outside the United States. Because the environmental advantages associated with more radical innovations may be substantially greater, some consideration of the differing factors is valuable. Generally, they can be viewed in terms of the structure and technology of the domestic paper industry, its integration with other major domestic industries, and the broad regulatory framework in which the issue has been viewed.

Although the U.S. paper industry is the largest in the world, in both Sweden and Canada the pulp and paper industry comprises a more significant proportion of the national economy. Importantly, in both countries the levels of shared and centralized research and of government/industry research collaboration are substantially higher than in the United States. Indeed, almost all the major U.S. competitors have stronger centralized research programs, all with government participation. In Canada, PAPRICAN (the Pulp and Paper Research Institute of Canada) is jointly funded by the federal government and the CPPA (Canadian Pulp and Paper Association) trade group and operates research facilities, projects, and degree programs in collaboration with Canadian universities.[37] In Sweden, much of the work on chlorine has been carried out by the SSVL (Swedish Forest Industries Water and Air Pollution Research Foundation) and similar groups. In the United States, however, despite the presence of various industry-funded research organizations (most significantly, the Institute of Paper Science and Technology, and NCASI, the National Council for Air and Stream Improvement, founded in 1943 but with an annual budget less than half that of PAPRICAN's in the early 1990s) and occasional collaboration between the USDA and various industry consortia and universities on special topics, no comparable organization conducts a broad range of centralized basic research and development activities for the paper industry.

Despite increasing discussion of the decline in scientists qualified in basic pulp and paper research in the United States, the continuing significance of industrywide environmental problems, and the fact that innovative technologies developed by one company may subsequently form the basis for technology-based regulatory mandates that affect all other companies, the domestic industry continues largely to be characterized by a focus on proprietary research efforts within large companies: it is far better organized in its lobbying efforts than it is around basic

R&D. Although the paper industry in all major producer countries has similar characteristics of scale, capitalization, and maturity, that generally tend to work against significant "natural" (or market-driven) innovation, the United States differs, in part, by virtue of lower levels and less effective forms of basic research capacity and by proprietary constraints on information sharing. The Agenda 2020 technology program discussed earlier is one recent response to this reality, but its effects have yet to be felt.

The close relationship between the paper and chemical industries is also of interest when considering the U.S. response to the issue of chlorine use in pulping. Both chlorine and sodium hydroxide (caustic soda) are produced as co-products from the electrolysis of an aqueous solution of sodium chloride (i.e., brine, or salt water), at a ratio of 1:1.1 (about 1 ton of chlorine is produced for every 1.1 tons of caustic). Both rank among the top ten commodity chemicals by volume of production. The chemical industry sector responsible for chlorine and caustic production is known as the chlor-alkali industry and is among the oldest and most mature sectors of the basic chemicals industry. Pulp mills have long consumed large quantities of both chemicals, with caustic soda central to both pulping and bleaching. In 1989, for example, the paper industry consumed about 1.8 million tons of chlorine and 2.9 million tons of caustic at a ratio of about 1:1.6. This reflected the first effects of increased chlorine dioxide substitution in the form of a rise from the traditional mill ratio (closer to the production ratio) of chlorine/caustic consumption.

As elemental chlorine consumption sunsets in the paper industry, caustic consumption within the industry is expected to remain at least stable, and perhaps to increase, with respect to pulp production levels. The rising imbalance in the domestic consumption of the two chemicals by the paper industry is broadly reflected across all industries as chlorine compounds in a wide variety of applications are targeted due to toxics concerns, resulting in dramatically fluctuating prices for the two chemicals, generally tending to higher prices for caustic and lower prices for chlorine.[38]

The U.S. chlor-alkali industry, with nearly a third of world production capacity (dominated in the United States by Dow Chemical and Oxychem), has responded in several ways. After dramatic expansions since the middle of the century, the past decade has seen the first declines in domestic chlorine demand. Some capacity has been shut down (in western logging regions), although some has been added (in petrochemical production regions) primarily for expanded PVC production. Overall, mid-1990s chlorine production levels were expected to remain fairly stable through the end of the decade, whereas caustic demand was expected to increase. Several companies have constructed facilities to produce caustic soda separately, by reacting sodium carbonate (soda ash) with lime.[39]

Not least, the industry has strongly mobilized to defend or expand its chlorine markets, originally in paper, but subsequently largely focused on PVC and exports. Efforts have been spearheaded by the Chlorine Chemistry Council, a Washington, D.C.-based advocacy group with an expected budget of $15 million in 1995.[40] In the United States, the chlor-alkali industry initially found a collaborator in the paper industry, which appeared substantially to have adopted the problem of resolving the chlorine/caustic imbalance in the face of dioxin problems as its own. As one reporter wrote, "the pulp and paper industry and its suppliers have recognized the impending caustic crisis and have been testing methods to alleviate the problem."[41] Although the paper industry had a clear motivation to retain the advantages of chlorine and to moderate the rising costs of caustic soda, the degree to which it shouldered the problem of chlorine/caustic imbalance seemed particularly pronounced in North America, especially relative to Europe. This was perhaps explained by the fact that although Western Europe accounts for a quarter of world chlor-alkali production, its paper industries consumed only about 4 percent of its chlorine output by 1990, whereas the figure was closer to 15 percent in the United States[42] As discussed below, reduced chlorine use in pulping began earlier and occurred more gradually in Europe.

For a time, between the late 1980s and the early 1990s, a flurry of activity focused on efforts to retain the use of chlorine in pulp production. Some sought to develop cost-effective on-site processes to convert chlorine (rather than the customary sodium chlorate feedstock) into chlorine dioxide.[43] Westvaco's multipoint chlorine injection process, which reduces dioxin formation but not chlorine use, was described as an example of "process changes to retain a paper bleaching market for chlorine."[44] Another analyst wrote, "with the help of the chlor-alkali industry, pulp producers are trying to retain chlorine's advantages while lowering dioxin levels."[45]

The characteristically low levels of R&D and the absence of cooperative mechanisms for technology diffusion created an underlying vacuum in the ability and inclination of the domestic industry to address the problems of organochlorine pollution from pulp bleaching. The various capital costs of process changes posed the next set of barriers to addressing the problem. In the United States, however, the more complex secondary effects of unbalanced chlorine and caustic consumption, combined with the influence of the chemical industry, appeared to strengthen the resistance to addressing the problem more comprehensively and rapidly. In a close pairing of the interests of the bleached kraft pulp industry with those of the chlor-alkali industry—both long-established, entrenched, mature industries—the problems of capital and scale, and the attendant structural barriers to change achieve truly formidable dimensions. Meanwhile, however, as European pulp producers continue a transition from ECF to TCF, a second generation of chemical industry activism around pulp bleaching issues has begun to emanate from the chlorate sector of the European chemical industry (sodium chlorate being the chemical purchased for on-site chlorine dioxide production, and thus central to ECF bleached pulp production, but eliminated in TCF).[46]

Another important difference between the United States and various European producers has been the regulatory context in which the problems of organochlorine pollution from pulp bleaching have developed. In Sweden, for example, the first commercial oxygen bleaching plant (now commonly viewed as an essential first step toward TCF kraft bleaching) was installed in the early 1970s—a decade before oxygen delignification first entered commercial use in the United States, and well before the pulp bleaching/dioxin problem broke on either continent. Sweden thus dates the beginning of its decline in chlorine use in pulp bleaching to about 1974 or almost fifteen years before the U.S. decline began. In Sweden, however, the first oxygen delignification systems were targeted at reducing conventional water quality problems created by bleach plant effluent, such as high organic load and dark coloration, and all kraft mills in Sweden have since installed oxygen delignification systems.[47] Reduced chlorine consumption was in this sense a fortuitous by-product of an upstream approach to solving wastewater problems that in the United States were primarily addressed by end-of-pipe treatment technologies. The development of oxygen delignification systems was stimulated by the enactment of the Swedish Environmental Protection law in 1969. Unlike U.S. pollution control laws, which developed independently around distinct media, the Swedish law covered releases to air, water, and land. Industrial environmental permits are not granted for an individual medium on the basis of binding emissions standards for individual pollutants; rather, a single integrated permit is granted by a national board on the basis of overall pollution levels, national emissions guidelines, and appropriate technologies.[48]

Although the need for integrated (across media) pollution control as a step toward a more coherent pollution prevention policy and practice has now become a prominent issue in the United States, Sweden clearly demonstrates both a longer tradition of integrated regulation and a greater capacity for encouraging fundamental innovation as a compliance response. It is also apparent, however, that although both Swedish industry and government were better situated (than their U.S. counterparts) to respond when the dioxin problem broke, the Swedish government chose to adopt a more aggressive and precautionary approach than that taken in the United States.[49]

In the late 1980s, for example, when Södra Cell sought to expand one of its mills near the Baltic Sea, proposing a high chlorine dioxide substitution process that yielded extremely low levels of both dioxin and total AOX, the Environment Protection Board nonetheless concluded:

> The Swedish EPB regard that today's discharges give an obvious (negative) impact on the receiving waters. The Swedish EPB regard that even if the observed disturbances are of relatively limited extent *they are danger signals that must be taken seriously.* The discharges must therefore be limited more than the company has promised . . . [50] (emphasis added)

In 1987 the EPB proposed an "Action Plan for Marine Pollution," which became the basis of a government bill subsequently passed by parliament, and stated, "In the long term, discharges of stable organic chlorine compounds from pulp and paper mills are to cease entirely." The Swedish approach was quickly followed by other countries in the region, and in 1990 the Nordic Ministers of the Environment agreed to the following chlorine bleaching policy: "The discharges of chlorinated organic substances should be eliminated altogether. It is now impossible to state the period of time or technology required to achieve this goal. The Nordic countries should aim at reaching this goal as soon as possible."[51]

As Ashford observes, "An industry's perception of the need to alter its technological course often precedes promulgation of a regulation." Indeed, as he points out, the preregulatory period of scrutiny of emerging regulations may be more important than the final rules because it stimulates and defines likely trajectories of innovation, whereas the final rule (typically expressed in technology specific forms) more commonly stimulates diffusion and technology transfer.[52] Although diffusion, or incremental innovation, may be an appropriate and desired response in some cases, radical technological innovations carry the potential to achieve far greater environmental benefits, often at an equivalent or lower per unit cost, especially when longer time frames and/or more comprehensive cost assessments are considered. The latter possibilities seem uniquely compelling in the case of the dioxin problem and in light of the persistence, severity, and extreme complexities of the multimedia pollution and risk problems associated with a broad range of organochlorine compounds and with their unmistakable connection to an identifiable source.

There is a certain boldness to a precautionary approach necessary if far-reaching gains are to be realized. And there are key moments in the trajectories of regulation and innovation—as difficult new problems begin to take shape, or as links between a set of problems begin to emerge—when this boldness can have the most effect. The initial show of concern by the government is thus of some importance. An important dimension of the response is the level of uncertainty that attends the preregulatory period:

> Uncertainty as to the ultimate regulatory requirement may be caused both by technical uncertainties and by the extent to which interested parties apply pressure for accommodation through both formal and informal means. But regulatory uncertainty is often necessary and beneficial. It is a necessary consequence of administrative flexibility, which allows regulations to be improved. Although too great a regulatory uncertainty may result in inaction on the part of industry until the outcome is definite, too much certainty about the final standard is not likely to result in the development of technology which exceeds the minimum requirements.[53]

In their initial responses, the Swedish and other Nordic governments conveyed both the certainty of a stringent and protective final goal (the elimination of organic chlorine compounds from effluent) and the flexibility embedded in the open timetables ("as soon as possible"). These policies were backed up by the development of a series of progressively more restrictive AOX and TOCl limits to be phased in. In Canada, the actions initiated by the social democratic B.C. environment minister to phase out all organic chlorine discharges from pulp mills have been followed in other provinces, partly reflecting escalating European demand for TCF paper and the importance of European exports to the Canadian industry (by late 1993 TCF had captured 60 percent of the German market for chemical pulp, and 15 percent of the western European market).[54]

In the United States, however, the initial regulatory show of concern is somewhat difficult to identify because it was buried almost from the outset under the interests and influence of the domestic industry, as first illustrated by the tainted collaboration between it and the EPA around dioxin studies in the 1980s. Despite the articulation of a national pollution prevention policy by the U.S. Congress in 1990, the institutional and statutory forces of traditional pollution control policies were still too entrenched to be significantly overcome in the pivotal preregulatory period of the mid- to late-1980s through about 1992. The quantitative risk-based approach to regulation, which has been forced to unprecedented extremes in the dioxin debates, obscured the potential of a more precautionary approach in which clear danger signals (distinct from quantified risk) must be taken seriously, as they have been in the Scandinavian countries and elsewhere in Europe and Canada.

Ironically, the early success of the domestic industry in squashing TCF-oriented regulatory tendencies has left certain large pulp manufacturers in an uneasy position with respect to the continuing European movement toward TCF, escalating costs of dioxin-related litigation and settlements, and already committed capital investment in chlorine dioxide substitution. Georgia-Pacific, for example, has on the one hand informed its customers that it will not meet TCF demand, but on the other hand claims that there *is* no authentic (consumer-driven) TCF demand. The latter refers to a sort of counterinsurgency theory that argues that countries like Sweden, Germany, and others simply found themselves well positioned to embrace TCF by virtue of historical providence (Sweden with its early move to oxygen delignification, Germany with its prohibition against kraft mills, and others because they are focused on mechanical and semichemical pulps). As the theory goes, these countries have embraced TCF-oriented regulatory mandates not because of long-range thinking or moral imperatives, but rather because their paper industries may gain a competitive edge.[55]

Difficulties with this theory include the fact that both Sweden's and Germany's lucky historical choices were in fact made not by accident, but on the basis of early recognition of environmental problems associated with the kraft process. To the extent that they gain a competitive edge on the TCF criterion, they probably deserve it. Also, British Columbia and Ontario, with the most aggressive standards of all (zero AOX), had no such historical luck. If the provincial standards hold, their large paper industries will face burdens that equal or exceed those now faced by U.S. manufacturers. With the number of provincial, national, regional, and European governmental bodies now involved, there is also a problem of international industry/government antichlorine and pro-domestic-paper collusion on a rather grand and risky scale, since numerous other major industries will also be directly or indirectly affected by stringent regulation of organochlorine emissions.

Defending its chlorine dioxide-based strategy and capital commitment (reportedly $100 million between 1992 and 1994), which has been called "a twenty-year investment for a ten-year technology," Georgia-Pacific has been most prominent among domestic producers fighting back with warnings that the "competitive position" of U.S. manufacturers will suffer if restrictive AOX standards are imposed and with the even more familiar claims that severe job displacement will be the result. "At 'no AOX' levels," Georgia-Pacific has asserted, "the number of displaced workers will rise to 36,000" due to the diversion of capital away from capacity expansion. The consequent multiplier effect—the displacement of related industries—may also be significant.[56] In a classic rebuttal, targeting a different multiplier effect, Mark Floegel of Greenpeace directly challenged Georgia-Pacific's economic impact claims during a panel discussion at the 1992 International Symposium on Pollution Prevention in the Manufacture of Pulp and Paper:

> The fact [is] that Georgia-Pacific has now lost two lawsuits in the state of Mississippi with total damages of, I believe, $3.25 million, because of discharging chlorine-based chemicals in the Leaf River Mississippi pulp mill. In addition to the two suits that have been lost already, 8,000 other pieces of litigation have been filed. Your insurance carrier, Aetna, has declined to pay for these damages and your own shareholders are suing Georgia-Pacific because you failed, allegedly, to file this litigation in your SEC [Securities and Exchange Commission] filings. These are real dollars and this does not even begin to count what's going on in the fishing industry in coastal America as a result of toxic pollution from a variety of industries. . . . That money represents income that has been lost, jobs that have been lost, hardships that have been endured in a monetary sense. We're not talking about the environment or health, we're just talking about the economy.[57]

Implicit in this exchange, and in the recent history of chlorine use in the paper industry, are the disjointed set of frameworks and boundaries that characterize the extraordinary difficulties of the transition from pollution control to pollution prevention and the associated progressions from quantitative risk-based approaches pushed beyond their inherent limitations to more timely and more precautionary strategies; from traditional narrow frameworks for capital budgeting and future costs estimation to more comprehensive assessments of costs; and from a focus on a deconstructed environment (a collection of media) in the immediate vicinity of a facility to a more interconnected ecological and social system.

In 1990, the EPA established a "pulp and paper cluster group" comprised of representatives from various media-specific (e.g., air and water) and other (e.g., policy, planning, and evaluation) EPA offices. Their charge was to integrate the process of addressing three separate and individually complex mandates: (1) to produce effluent guidelines for dioxin releases under the Clean Water Act; (2) to define MACT (maximum achievable control technology) for hazardous air pollutants released by chemical pulp mills under the Clean Air Act amendments of 1990; and (3) to regulate dioxin and furans in the land application of sludge under TSCA. The latter was eventually pursued separately from the development of the air and water cluster rules.[58]

By at least as early as February 1993 it was clear that the EPA would not pursue a TCF mandate for kraft mills.[59] Nevertheless, the Phase I cluster rules proposed in the Federal Register on 17 December 1993 represented what one analyst described as "a compliance nightmare or godsend regulation (depending on your viewpoint)."[60] Under the whole plant effluent category new pollutants including AOX, color, chemical oxygen demand (COD) and dioxin were added, and limits for conventional pollutants such as BOD and total suspended solids (TSS) were tightened by as much as 90 percent. Bleach plant effluent was regulated separately (i.e., upstream of treatment and release), with fifteen specific compounds or classes subject to best available technology (BAT) requirements including TCDD, TCDF, chloroform, acetone, methyl ethyl ketone, methylene chloride, and various chlorinated phenolic compounds. The effluent controls were expected to require extended digester delignification and/or oxygen delignification combined with 100 percent chlorine dioxide substitution for bleached softwood kraft mills—that is, ECF.

Separate standards for paper-grade sulfite mills, however, would require them to practice TCF bleaching. New "brown grade" (nondeinking) wastepaper-based pulp lines would be required to be totally effluent free, and various other categories of mechanical, semichemical, and dissolving woodpulp mills would be affected to varying degrees. Not surprisingly, however, the focus has been on kraft pulp mills. Under the Clean Air Act portion, mills are required to meet MACT standards for some fifty hazardous air pollutants (HAPs) for which the industry was identified as a major source, prominently including methanol, chlorine, hydrochloric acid, chloroform, toluene, and formaldehyde, among others now famous from TRI reports.

Studies sponsored by the AFPA suggested cluster rule compliance costs for a representative 1,000 tpd bleached softwood kraft mill could range from $53 million to more than $220 million. The EPA estimated total compliance costs to the industry at $4 billion, whereas the AFPA set the figure at $11.5 billion and further suggested the likelihood of multiple mill closures. In either event, these expenditures come on top of the $4 billion estimated to have been spent by the industry between 1988 and 1992 on reducing dioxin formation, improving wastewater treatment, and reducing kraft mill odor problems.[61]

Pollution Prevention from Other Perspectives

The final cluster rules have yet to be promulgated as this is being written, but whether eventual compliance costs total $4 billion, $12 billion, or some other number, if promulgated in a form similar to the 1993 proposal the rules will likely involve unprecedented levels of capital expenditure over a short period of time in the name of environmental improvement. In this light, it is worth considering what the impact of these rules may be from various other perspectives.

What, for example, will be the effects of the drain on both human and financial capital these regulations imply?[62] If already-limited technical resources are channeled to cluster rule compliance projects, what will their availability be for more long-range investigation of future technologies? Will capital that might be channeled into wastepaper-based production instead be channeled into kraft mill upgrades, or will it be enhanced if some chemical pulp mills are forced to close? If mills are forced to close, will they more likely be small- or medium-sized—or large-sized? If the former, what does this say about the already high concentration and dominance of large-scale formats in the industry, and the myriad negative effects of associated structural rigidity? As mills spend tens or hundreds of millions of dollars on capital programs, will the pressure to reduce labor costs even further accelerate the already negative employment trends? Does the ECF focus of the cluster rules with respect to kraft mills stall or undercut technology that might have leapfrogged to TCF, closed-cycle bleach capability, with a similar level of investment? Or do the rules for paper-grade sulfite mills, which would require them to practice TCF bleaching, in fact set a "dangerous precedent" as some industry analysts have suggested? Are things looking up or down for the paper chemicals divisions of large chemical manufacturers? What will the impact on wood resources be? Paper imports? Pulp exports?

Although educated guesses can be made as to some of the answers, the underlying point these questions raise is the significant degree to which such regulations can affect the structure and trajectory of the entire industry. In turn, the direction taken by the industry raises implications that range far beyond the pollution issues addressed in the regulations. Because of the degree of capital investment likely to be committed, they chart a course that may endure for some time to come. Inadvertently or otherwise, the cluster rules in these respects constitute a characteristically murky and unspoken form of domestic industrial policy.

The cluster rules may challenge the industry, but one can argue that they do so essentially by forcing it to play catch-up with other countries and with domestic laws that have been on the books but stalemated by litigation and industrial opposition for decades (the regulation of hazardous air pollutants being a salient example). The challenging part essentially comes from the fact that the industry must catch up all at once. In this respect, the new technology research partnership with the industry could be read as something of a salve for whatever wounds it may sustain, notwithstanding the industry's own responsibility for its loss of technological leadership or its generally reflexive opposition to regulation. Yet, it could also be read as a serious attempt to begin to redress the industry's underlying deficiencies in R&D.

Another interesting aspect of the cluster rules is that although they advance the general concepts of integrated pollution control, in-plant versus end-of-pipe regulation (by specifically regulating bleach plant effluent), and more holistic industrial regulatory approaches, they do so with baby steps. Using Frances Irwins's distinction, the *internal* integration of the single-medium approach has begun, but the *external* integration with other policy areas has not.[63] In this respect, the EPA's Common Sense Initiative, which builds on the cross-medium, sector-focused cluster rule model, also remains handicapped, although at the same time representing progress. The Common Sense Initiative, announced in early 1994, establishes teams of industry-sector stakeholders from the target industry, the EPA, and national and grassroots environmental organizations for each of the six industries selected (which include steel and petroleum refining). The groups are charged with examining pollution control and prevention strategies and regulations, and with identifying opportunities to streamline, integrate, and improve them on an industry-wide basis.[64] The limitations of the approach are largely found in the limitations of the EPA's mandate, which essentially begins at the factory with pollution and proceeds from there.

As suggested in the preceding chapter, however, a materials policy perspective recognizes that changes in materials use at the front and middle stages of the economy may be equally germane to pollution prevention opportunities. As with wastepaper in the last decade, the now-emerging argument for the comparative pollution prevention benefits of nonwood pulping is particularly relevant, although it has yet to be thoroughly demonstrated by either basic research or commercial experience. Distinct from resource-focused arguments advanced in support of increased nonwood fiber use, the foundation of the manufacturing-focused pollution prevention argument for nonwoods is generally the lower average lignin content of nonwoods, combined with other less well characterized differences in the chemical composition and cellular structure of annual plants (such as a relatively porous structure more easily penetrated by chemicals). Although offering relatively few details on the comparative pulping responsiveness of selected nonwoods versus benchmark wood types, the extensive nonwood pulping studies by USDA researchers in the 1960s generally concluded that annual plants "usually have lower lignin content and higher hemicellulose content than most woods, and they show extremely rapid response to refining . . . [T]his property is so marked that investigators accustomed to handling wood pulps are likely to subject annual crop pulps to conditions more rigorous than optimum."[65]

The lignin content of wood falls in a range of roughly 26–34 percent (softwoods) and 23–30 percent (hardwoods), whereas for nonwood furnishes the range is estimated to be at least as wide as 3–31 percent. For nonwood furnishes that have lately been discussed in terms of commercial potential in the United States the following general ranges in lignin content have been suggested: kenaf whole stalk (14–22), bast (8–17), core (17–22); hemp bast (3–4), core (20–22); and various straws (10–20).[66]

Because the breakdown and/or extraction of lignin is the core objective in all pulping processes whether mechanical or chemical, and in all chemical pulp bleaching processes, beginning with a material that has comparable cellulose content, but 30–50+ percent less lignin (and/or more easily extractable lignin), offers obvious potential in terms of reduced energy and chemical use. TMP processes, for example, have been demonstrated to produce good quality, high-yield whole stalk kenaf pulps with up to 25 percent less energy used in refining than reference southern pine furnishes—gains that can be even further enhanced in biomechanical and CTMP configurations.[67] In general, the technological feasibility, energy advantages, and acceptability of pulp qualities offered by a variety of nonwoods appear to be unchallenged in the case of mechanical pulps (although fiber security issues, strongly connected to the scale of existing mills, continue to be raised).

The basic chemical processes available to pulp nonwoods are the same as those used to pulp wood. Because the primary objective is delignification without damage to cellulose fiber, lower lignin content is considered to offer a significant advantage here as well (for example in cooking time, chemical charge, or temperature relative to pulp yield, delignification, brightness, and strength). Unfortunately, comparative (wood versus nonwood) pulping responsiveness and associated environmental profile data (chemical use, energy and water use, emissions factors, and yield) under constant pulping conditions, combined with characterization of the associated pulp, are almost completely unavailable in published western literature. The early USDA chemical pulping studies (using kraft, soda, and neutral sulfite processes) examined the comparative pulping responsiveness of various annual plants to one another (not to wood), although the obtained pulp characteristics were thoroughly compared to reference wood pulps, and the ease of refining was noted. It is not clear to what extent data based on these studies can be compared to modern implementations of the same basic processes. However, the parts of the picture that can be pieced together from recent evidence tend to confirm the significant potential of many nonwood furnishes in terms of reduced processing. That it is possible, under conventional chemical pulping conditions, to produce nonwood pulps with properties that compare favorably with both hardwood and softwood pulp qualities has long been recognized.

Al Wong reports that a sisal furnish (which has an average lignin content of only 8–9 percent) can consistently produce a very strong pulp with an *unbleached* brightness of 85 points in a standard alkaline sulfite process (bleachable to 89 points with a single-stage hydrogen peroxide treatment). He makes a more general case for the use of modern sulfite processes in nonwood applications due to their ability to achieve high brightness with little or no bleaching and to the comparative flexibility of sulfite in whole-stalk pulping of bast fiber plants such as kenaf, relative to the kraft process, which is not well adapted to cooking fibrous raw material with different lignin content at the same time.[68]

KP Products of New Mexico now produces a high-strength, premium quality, uncoated kenaf printing paper in a semi-bleached "natural" color (72 points brightness). It is based on a blend of preseparated kraft-pulped kenaf bast and core fibers, bleached in a single stage peroxide process with a chelant pretreatment (QP). Spent pulping chemicals are evaporated and shipped to an off-site kraft mill for recovery.[69]

Generally, strong, long-fibered nonwood pulps such as those made from abaca, sisal, kenaf bast, crotalaria, and bamboo are viewed as partial or total substitutes for high-grade softwood pulps; and weaker, short-fibered nonwood pulps from various straws, core fibers, and bagasse are viewed as substitutes for hardwood pulps.[70]

Unlike mechanical nonwood pulping, however, technical barriers remain associated with the conventional chemical recovery process for full chemical pulping of nonwoods, especially straws, in the inorganic regimes (kraft, soda, sulfite). Nonwoods generally have a higher ash (inorganic compounds) content than wood, of which a primary constituent is commonly silica, although the variation among different plants in both ash and silica content is extremely high. Total ash content may range from below 1 percent up to about 25 percent in nonwoods, and silica may range from less than 1 percent up to as much as 90 percent of ash content (or from negligible up to about 15 percent of total plant composition). For some plants such as sisal, total ash content is low and more comparable to wood (<1 percent), whereas for most straws it is high (roughly 5–25 percent).

The ash content of kenaf ranges from about 2–6 percent (lower in bast, higher in core), roughly comparable to hemp, bamboo, and bagasse. Total silica content, negligible in wood and only slightly higher in sisal and kenaf, ranges from about 3–15 percent for different straws and is particularly high in rice straw. Most straws also have comparatively higher potassium content than other plant fibers.[71] Both silica and potassium are nonprocess elements that are difficult to precipitate out of the black liquor in conventional inorganic chemical recovery processes. Because up to 99 percent of inorganic chemicals present in black liquor are recovered in high-efficiency modern recovery systems, these nonprocess elements, if significantly present in the furnish, can eventually build up to levels that interfere with operation of the recovery system.

The chief reason that most experience with nonwood chemical pulping as practiced in developing countries (primarily based on straw, bagasse, and bamboo) cannot be easily translated to the United States or Europe is that environmental regulatory controls are substantially less developed, thus no recovery or partial recovery of black liquor is still practiced. (It is also one of the reasons these mills are generally associated with severe levels of pollution.) Research efforts have focused largely on improved processes for extracting silica from the spent kraft cooking liquor, although the potential for shifting from sodium-based to potassium-based processes (e.g., from sodium hydroxide to potassium hydroxide as the basic digester chemical), has also been suggested. The variation in technical and economic feasibility of different approaches may be high depending on differing fiber characteristics, process configurations and capacities at existing mills, chemical costs, and so on; however, basic R&D is severely lacking in the United States. The problem appears most serious in the case of agricultural residues, but may also be a factor in the use of other nonwoods, which have lower inorganic content than most residues, but still higher than wood.

Showing promise in the longer term are organic solvent pulping regimes. Their simple recovery systems appear to be well adapted to the recovery of alcohol solvents, lignin and other co-products without being negatively affected by inorganic residuals. Their smaller economies of scale (about 300 tpd) are also better adapted to the smaller scales considered appropriate to nonwood pulping for fiber supply reasons. As mentioned earlier, however, solvent pulping has never been commercialized in the United States, and current research on it is concentrated in other countries. As with other chemical processes, no thorough comparative analysis of wood-based versus nonwood-based organosolve pulping could be identified. However, because hardwood-based solvent pulps can achieve high brightness in TCF bleaching, it is expected that many nonwood furnishes would be equal or superior in this regard. Dutch research on ethanol-based pulping at lab scales has demonstrated effective pulping of preseparated hemp bast and core fibers, with various papermaking applications possible, but more research is required.[72]

Representing an entirely different approach, Al Wong has patented a potassium-based alkaline sulfite pulping and recovery process that appears particularly well adapted to the typically high levels of potassium and silica present in agricultural residues such as wheat and rice straw. The primary pulping chemicals are potassium hydroxide and potassium sulfite. Chelant/peroxide (QP) bleached brightness of wheat straw pulped in this process is about 84 points. The black liquor consists chiefly of the potassium chemicals (now mostly in the form of potassium sulfate and potassium carbonate) and various lignin by-products dissolved in water. It is evaporated (with vapor recovery/reuse) to 65 percent water and without further processing constitutes a marketable fertilizer, consisting mostly of potassium compounds and lignin by-products. In this totally effluent free process, recovery is thus essentially limited to containing and evaporating the spent liquor, which is then shipped as a product in liquid form. As with the sale of the lignin co-product in pure solvent pulping, the sale of the fertilizer co-product is integral to the economics of this process. Due to the simplicity of the process it is expected to be feasible at very small scales (as little as 20,000 tpy or 50–60 tpd) and is thus uniquely well adapted to fiber supply issues associated with agricultural residues. A pilot-scale version of this process is now operating in the province of Alberta.[73]

A similar process based on an ammonium bisulfite cooking liquor has been implemented in some twenty-six small mills in China. Although the process is still being perfected, the spent pulping liquor is being successfully used as a value-added fertilizer.[74]

A Question of Strategic Vision

When one steps back from the minutia of organochlorine and other pollution problems and of their technological remedies in kraft pulping, one begins to see the serious underlying problems associated with the dominance of kraft woodpulping in the United States. Not least are the huge production scales it has come to entail as well as the low levels of innovation and flexibility that have been one of the prices associated with these vast economies of mass production. One also sees that many of the under-investigated nonkraft remedies to organochlorine pollution (e.g., solvent and modern sulfite pulping) seem uniquely well adapted to nonwood pulping, which in turn may well offer additional pollution prevention as well as resource conservation opportunities at both pulping and cultivation levels. It also becomes clear that several of these nonkraft technologies most favorable from both TCF and nonwoods perspectives are also favorable for smaller pulping scales.

Recently, a few research efforts have been proposed in the United States that would begin the process of characterizing the comparative impacts of fiber cultivation, preparation, and transportation, and of chemical pulping, recovery, and bleaching under different regional, crop (wood, nonwood, residue), and process scenarios. They have been framed in terms of goals as diverse as identifying opportunities for chemical use reduction in the agricultural fiber and pulp industries, averting fiber demand crises in northwestern regions, and reducing greenhouse gas emissions. Prospective federal funding sources have ranged from the Department of Energy to the USDA and the EPA.[75] It is safe to say, however, that the current resources available for these efforts are vanishingly small in comparison to the corporate and governmental resources now committed to conventional woodpulping and fiber supply.

The solutions suggested by the modern EPA pollution prevention perspective, or a kraft mill perspective, are not necessarily the same as those suggested when one takes a longer-term view, or when one backs up enough to include *both* pollution prevention and sustainable resource use in the field of vision, or when one considers the prospective social and economic advantages of a more decentralized industry. Indeed, the conventional solutions may strongly contradict the significant opportunities that arise from the more comprehensive viewpoint. Although the cluster rules have yet to be finalized, their development has, predictably, already begun to stimulate and direct capital investments in the industry. From the more comprehensive perspective, the signs are not encouraging. A number of bleached kraft mills have recently undertaken major capital investment programs, adding oxygen delignification systems and making other modifications to bleach plants, digesters, or recovery systems; in some cases total production capacity has been simultaneously expanded to levels approaching a million tons per year.[76] Even if capacity is not expanded, the reinvestment in these large, wood-based plants gives them a new 20–30 year lease on life. In this way, the rules may both update and reinforce the current organization of the industry, and thus further reinforce barriers to promising nonwood and nonkraft alternatives.

One might think it desirable for regulators—who have the power to establish regulations so significant that they function like a de facto industrial policy—to be more far-reaching in their vision. Yet, because they are still digging out from under the burdens of existing medium-specific and other mandates, attempting to integrate, rationalize, and modernize them (all the while facing challenges to their existing authority), it would be at least naive to pin high hopes in this direction in the near term. These points do tend to indicate, however, that proactive sector-focused pollution prevention, as a federal or state policy, will ultimately require the participation and support of other government sectors at a far more profound level than pollution control approaches have. The real subject matter of pollution prevention extends well beyond the institutionalized boundaries of technology and industry observed by the still-dominant pollution control framework and crosses over into the marketplace and into systems of product use and raw materials choices long considered outside the bounds of pollution control.

Of course, neither the industry nor the larger environmental community must wait for regulators to chart the course. Environmentalists have the unique responsibility of driving the discourse and putting new elements on the agenda. Yet, the environmental community itself continues to be fractured along fault lines that in part mirror the divisions found between and within regulatory agencies. The progression in focus of Earth Island's paper campaign—from tree-free paper to forest friendly paper to ReThink Paper—is symptomatic (although the rapid evolution is encouraging). The term *tree-free* troubled recycling advocates by seeming to discriminate against their objectives. Indeed, the potential for shallow divisiveness on this point has been leveraged by the established paper industry.[77] The term *forest friendly* solved this problem, although it did not overtly reflect any commitment to pollution prevention—particularly at issue because of the severe pollution associated with nonwood pulping in China and elsewhere. At the same time, however, one can also question whether antichlorine campaigners operating at national policy levels have fully considered the degree to which a narrowly defined agenda might act to reinforce a large-format industrial system inextricably tied to the use of wood by driving huge capital reinvestment in kraft woodpulping.[78] Mainstream sectors of the environmental community continue to be preoccupied with optimizing the status quo, with only incremental improvement, and offer no real challenge to—and thereby tend to reinforce—the fundamental structural problems in the industry, which underlie all its environmental problems.[79] The notion of *rethinking paper* certainly enlarges the domain, but one can further argue that what is really needed is to rethink not just the products (or the chemicals, or the technology, or the resource base) but the industry itself.

The industry itself is of course not without its own opportunities. Although the established industry is the best situated to pursue alternative courses (by virtue of its immense resources and concentrated power), it is simultaneously handicapped by its existing commitments to wood and its narrow charter (maximize profit). As USDA kenaf researchers recognized decades ago, "[K]enaf's economic potential resides in the industry's strategy adopted to cope with the future," and the industry today continues to look to wood.[80] Nevertheless, some surprising things have lately begun to show up in the trade literature. One is dissention in the ranks, symbolized by the TCF/ECF schism, and by recent defections from AFPA membership (apparently reflecting, at least in part, a split between the integrated forest products companies and those focused more exclusively on paper products).[81]

An even wider schism has continued to develop between the U.S. and European industries in terms of long-term perspectives or lack thereof. Despite the shocks to which it has lately been subjected, the core sectors of the domestic industry continue to hold to a vision of the future apparently constructed by looking backwards, or at least, as Eklund claims, no further ahead than the quarterly results or the next board meeting. While the domestic industry cannot help but have absorbed the reality that its environmental problems go far beyond a temporary public relations glitch, it has been unable to accept that this implies the need for fundamental change in its long-range production strategies.

Consider the unabashed glee with which AFPA President W. Henson Moore greeted the virulently antienvironmental Republican majority elected to the U.S. Congress in 1994. Writing of the paper industry's high expectations for a turnaround on the regulatory front, he stated:

Industry is not alone in sharing this expectation, of course. Today it finds new intellectual allies from academia questioning the economic efficiency of the command and control *status quo*. And new political allies, too, from grassroots groups indignant over the loss of private property rights, to local governments complaining of unfunded mandates. In fact, fueling congressional calls for a panoply of reforms—for ending unfunded mandates, supporting regulatory moratoriums and risk assessment, and the devolution of regulatory authority to the states—are insistent complaints from state and local governments about the rising cost of environmental regulations.[82]

Perhaps most remarkable for the thinly veiled embrace of some of the furthest reaches of militant ultraconservativism, these comments stand in sharp contrast to perspectives now coming from the European industry. For example, Stefan Kay, chairman of the Paper Federation of Great Britain, openly acknowledges that today's pulp mill, "large-scale, high-technology, immensely costly," may already be obsolete and observes that "increasingly stringent environmental rules will force massive expenditure with no direct benefits to the bottom line, apart from the ability to stay in business." Instead of attempting to deny this eventuality, he looks to alternative feedstocks, and alternative pulp technologies that lend themselves to "smaller scales, less pollution, and lower capital costs."[83]

The basic economic irrationality of wholesale resistance to increasing environmental pressure is similarly targeted by Dutch analysts Mark Weintraub, Hans Grüfeld and Pieter Winsemius, who state:

> . . . [T]he industry's response to the environmental challenge has been far from adequate. In environmental terms the response has been slow and narrow in scope. As far as shareholder value is concerned, it has in most cases been detrimental in both the short and the long term. It has ignored market opportunities and competitive positioning, and typically been piecemeal, late and, as a result, costly.[84]

Overall, these and other European analysts and industrialists have begun to argue for a "smarter response" based on thinking beyond compliance and on integrating the environmental challenge into a long-range strategic vision, one that may imply a "major break from current trends." On the question of strategic vision, one can conclude that both the established domestic industry and the environmental community itself are well advised to pay attention.

4

Paper Recycling

Recycling rates have received a lot of attention since municipal solid waste management emerged as a serious national problem in the 1980s. What has often received less attention, however, is what we actually mean when we talk about recycling rates, and whether the various things we mean really tell us what we need to know. The lexicon of modern recycling policy and analysis is complicated and still fluid. To some extent this is to be expected of any policy area that has emerged and evolved rapidly. To a large degree, however, much of the confusion can be traced back to the overwhelming dominance of the municipal solid waste view of recycling. This framework has become so deeply entrenched that it has, in a sense, come to "own" the potential of recycling. This chapter argues that the utility of this framework has long been limited and that broader perspectives are needed to advance the real potential of intensive recovered materials use: resource conservation, pollution prevention, and waste reduction. The following sections outline both conventional and unconventional views of domestic progress on the paper recycling front between the mid-1980s and the early 1990s, and begin by considering the vocabularies of recycling rates, recovered fiber grades, and paper products.

Recycling Rates and Fiber Utilization

The first comprehensive national characterization of municipal solid waste (MSW)—a seminal document for the solid waste and recycling developments of ensuing years—was prepared by Franklin Associates for the EPA in 1986. This study disclosed an overall *national MSW recycling rate* estimated at about 10 percent, updated to 13 percent in 1988 and to a little over 14 percent in 1990.[1] The figure is calculated as tons MSW *recovered* for recycling, divided by tons MSW *generated,* but it is somewhat more noteworthy for what it leaves out of the definition of MSW, than for what it includes, and for the difficulties in describing the difference. The ambiguity derives largely from problems inherent in clearly distinguishing between various municipal, industrial, commercial, institutional, and household wastes—and their sources.

Solid waste terminology, particularly the distinction between *industrial solid waste* and *municipal solid waste,* has a complex etiology that reaches back to early federal involvement in solid waste management in the 1950s and 1960s. The distinction between the two was first codified in the Resource Conservation and Recovery Act of 1976 (RCRA), which addressed the "third pollution," solid waste disposed on land, following earlier regulation of emissions to air and water. It was thereafter voluminously elaborated in ensuing regulations and amendments. Subtitle C of RCRA addressed *hazardous waste,* which came to be defined through long lists of hazardous characteristics, chemical constituents, and associated concentration thresholds of concern; related lists of industry and process-specific waste streams; volumes of facility and waste-specific exemptions; and various state lists that augmented all the preceding. Taken together, they came to constitute a tortured and effectively incomprehensible definition of the "solid" (or at least containable) fraction of toxic and/or otherwise dangerous solid, semisolid, liquid, or gaseous—and not already regulated—industrial wastes. Subtitle D addressed what we have come to call variously municipal solid waste; residential, commercial and institutional waste; or *postconsumer waste.*

Andrew Szasz has argued, in an analysis of the legislative history of RCRA, that a strong, comprehensive argument for "waste reduction at the source" was made in those debates. Yet, appearing to open the door to overly direct government intervention in the production process, this generic argument for waste reduction became a factor in the ultimately strict segregation of industrial and municipal waste management policies. One of the bills introduced by Democratic Party legislators, for example, had suggested that "standards may include minimum percentages . . . maximum permissible quantities of component materials and may prescribe methods of distribution . . . and prohibitions against the manufacture and sale of specific items." These were not the procurement standards for recycled products seen in more recent years, but rather production standards that were being proposed. The eventual finessing of solid waste into hazardous and postconsumer (or industrial and municipal) categories could in part be seen to reflect a broader effort to channel these aggressive tendencies away from producers and toward the consumer and postconsumer waste where they could be more safely debated. As Szasz writes, "corporate opposition to source reduction for postconsumer waste served as a kind of proxy through which [industrial waste] generators moved to squelch any incipient notion of addressing producer waste through government regulation of production."[2]

The ramifications of this political differentiation between industrial and municipal waste were extensive and underlie much of what has become most confusing about the modern recycling vernacular and most confounding in pollution prevention terminology. The RCRA classification scheme for hazardous wastes, for example, has evolved into what has long been described as "one of the most intricate and bewildering frameworks that exist in any area of federal law."[3] Databases developed from the reporting of RCRA-defined hazardous waste generation and management have, in turn, defied even the most valiant attempts to chart waste reduction, waste minimization, and pollution prevention progress meaningfully. (Not surprisingly, such attempts have been largely abandoned since the less-bewildering TRI data became available as an alternative indicator.)

The obvious commonalities in material properties and waste management options associated with nonhazardous producer wastes (such as botched publishing runs) and bona fide end-user wastes (such as household wastepaper) have been artificially complicated by efforts to construct bullet-proof definitions of postconsumer waste. The *hazardous* fraction of postconsumer wastes (like the *nonhazardous* fraction of industrial wastes, a confusing tautology in the official schema) has, of course, been largely unmanageable in isolation from its industrial origins. When we speak of source reduction of postconsumer waste (or MSW) within this framework, we have difficulty pinpointing the source. It is unclear if we speak only of office workers who print double-sided copies or if we mean to include copier manufacturers who make that easier to do or paper manufacturers who lower the basis weights of their products. Is source reduction of postconsumer waste something that only consumers can do, or can producers do it too?

Overall, the logical continuum between producers, consumers, and wastes has been shattered, although more recently, as source reduction of postconsumer waste has been taken up more earnestly, it has begun to be pieced back together. However, constructs like the *MSW recycling rate* and *postconsumer waste* are grounded in these imprecisions and conflicts with respect to industrial and consumer sources of waste, and they ultimately encourage the separation of the consequences of production from the consequences of consumption. They are thus, at best, an incomplete and sometimes misleading counterpart to what must necessarily be a more useful, reproducible, and comprehensive set of metrics for the analysis of recycling progress.

In a 1973 analysis of domestic paper recycling for the American Paper Institute, William Franklin (of Franklin Associates, the consulting group that has dominated the field of public and private analysis of MSW in general and paper recycling in particular for more than two decades) defined the paper recycling rate as "the quantity of wastepaper used as a raw material in paper and paperboard in any time period divided by total national consumption of paper and paperboard" (or tons wastepaper *utilized* in production divided by tons paper *consumed*).[4] This early, hybrid figure illustrates two variations on the MSW recycling rate described above. The first is that recovered materials utilized in production (limited to wastepaper in this case) are differentiated from total recovered materials. The second is the use of total paper *consumption* as a denominator rather than municipal *waste generation.*

The latter distinction is somewhat subtle, but, generally speaking, paper consumption (or *apparent consumption:* total production, minus exports, plus imports) can be more easily calculated at the national level because it uses actual commodity production and trade reports data that are widely available and well established. Actual wastepaper generation is not widely reported with any consistency and thus must be estimated based on various assumptions about consumption and about the definition of MSW. The use of apparent paper consumption as a proxy for wastepaper generation assumes that the amount of paper produced in the country, plus net imports of paper commodities, equals the total amount of paper that will eventually be disposed of (or recovered) in the country. As a platform for discussing wastepaper generation, recovery, and so on, it has the advantage of including the postmill/preconsumer commercial and industrial wastes that result primarily from converting and publishing operations, which are not commonly included in the calculation of MSW (although they constitute about a quarter of all wastepaper recovered for use as of the early 1990s). It does not, however, include hidden paper flows associated primarily with packaging. For example, paper that enters or leaves the country as packaging for "cantaloupe exports" will not show up in commodity paper consumption and trade figures, although it would be reflected in actual wastepaper generation data if it were available. Paper commodity trade data is also not available at the subnational (i.e., interstate trade) level.

As reporting systems attempting to chart actual waste generation at the state levels have developed, however, they have sometimes provided an interesting glimpse into regional patterns of paper flow. In 1990, for example, apparent per capita paper consumption was about 700 lbs in the U.S. Figures developed from municipal and county reporting requirements under California's solid waste regulation efforts, however, indicated that more than 14 million tons of wastepaper was generated in the state in that year, or more than 900 lbs per capita.[5]

The potentially recoverable portion of total apparent paper consumption includes paper that is not composted at residences, flushed into sewers, or otherwise heavily contaminated or destroyed in use. It is certainly more than 50 percent of consumption given the current mix of products, and different estimates of recovery potential have ranged as high as about 85 percent of consumption.[6]

In 1989 the congressional Office of Technology Assessment (OTA) took the broad road by defining the *recycling/recovery/diversion rate* as "the tonnage of recyclables collected and processed into new products divided by the total tonnage of MSW generated." This figure is similar to the MSW recycling rate used by the EPA (in using estimated MSW generation as the basis), but, like the early paper recycling rate calculated by Franklin, the OTA figure also indicated that not just recovery, but also the utilization of recovered materials defined recycling.[7] These different definitions have underlain the distinctly different sets of numbers used to chart recycling progress. In 1990, for example, the EPA estimated that 20.9 million tons of (postconsumer) wastepaper were recovered, for a recovery rate (relative to MSW) of 28.6 percent. By contrast, the American Paper Institute estimated that 29 million tons of (total) wastepaper were recovered in the same year for a recovery rate (relative to apparent consumption) of 33.6 percent.[8] Both definitions, however, advanced the approach begun by the OTA in differentiating between the ambiguous term *recycling rate* and the more accurate term *recovery rate,* although they did so by different routes. The definition of recovery rate based on total paper consumption and total wastepaper recovery is used in this chapter.

However, recovery rates standing alone as the measure of recycling progress are grounded in a waste-management view of recycling because they do not disclose anything about what happens to the wastepaper that is recovered. If, for example, one is interested in the forest protection, energy conservation, or pollution prevention advantages of recycling, it is also necessary to know a number of other things, including whether the wastepaper is being used in domestic paper production and whether it is reducing virgin materials use.

Unlike wood and nonwood fiber-based pulp production comparisons, which are still at an early stage of development, the intuitively obvious pollution prevention and energy conservation case for wastepaper-based paper production (versus both waste management alternatives and wood-based production) is by now thoroughly updated from 1970s analyses, and extensively documented. The Environmental Defense Fund and its corporate partners on the Paper Task Force have most recently argued the case in the process of developing recommendations for purchasing "environmentally preferable" paper—"paper that reduces environmental impacts while meeting business needs."[9] As discussed later in this chapter, the EPA also strongly made the case in the late 1980s in proposing, under the Clean Air Act, source separation and recycling requirements for wastestreams in municipal incinerator service areas. The potential for fostering energy conservation through recycling has long been so apparent that it was specifically included in the statute that created the U.S. Department of Energy in 1977.[10]

A quarter century of analysis and common sense notwithstanding, various surprising arguments continue to be advanced from some corners against the pollution prevention and energy conservation comparison of wastepaper and wood-based production. A recent, extraordinary example is found in a 1994 report prepared for the FAO by Jaakko Pöyry Consulting.[11] Examining high and low wastepaper utilization scenarios in four countries (Sweden, Germany, Romania and the Philippines) the report concluded: "When total environmental effects are considered there is no doubt that the *scenarios with the lower consumption of recycled fibre for paper production are the most positive ones.*" It ended with the even more astonishing observation that "public opinion must be changed," lest future recycling development prove "disastrous for paper, for the paper industry, and for forestry." Such conclusions can only be attributed to what must surely be one of the worst abuses of life-cycle analysis ever published.[12] Nevertheless, they give some sense of the level of resistance that has had to be overcome, and of the role that dogged common sense has played in the domestic developments discussed below.

Two figures are important in evaluating the degree to which wastepaper is being not just recovered, but also used: the *wastepaper utilization rate,* and the *wastepaper export rate.* The wastepaper utilization rate is calculated as: tons wastepaper utilized in paper production divided by tons paper produced. The export rate (or the net export rate) can be calculated as: tons (net) wastepaper exports divided by tons wastepaper recovered. Imports of wastepaper are negligible, so the wastepaper export rate is effectively equal to the net export rate.

With these definitions, it is possible to appraise both recent progress in recycling and the historical context in which it should be viewed. According to the AFPA, the wastepaper recovery rate increased slowly but steadily from 22 percent of consumption in 1970 to almost 27 percent in 1980. The recovery rate stalled after 1980, although by the mid-1980s the effects of the latest recycling movement could be seen as the recovery rate shifted upward and increased to almost 32 percent by the end of the decade, and to 38 percent by 1992 (see table 4.1).[13] This exceeds the wastepaper recovery rate achieved by the U.K. and France by the same time (32 and 36 percent, respectively), but lags behind countries such as Germany (54 percent) and the Netherlands (63 percent).[14]

Table 4.1
Wastepaper recovery, export, and utilization, 1970–2000 (million tons)

	Paper and paperboard production	Apparent paper and paperboard consumption	Consumption including converted products	Waste-paper recovery	Net waste-paper exports	Waste-paper used*	Recovery rate (% cons.)	Export rate (% recovery)	Utilization rate (% prod.)
1970	51.7	56.1	56.0	12.6	0.5	12.0	22.4	4.3	23.3
1971	53.2	57.5	57.5	12.9	0.6	12.3	22.4	4.5	23.2
1972	57.4	61.6	62.0	13.7	0.6	13.1	22.1	4.1	22.9
1973	59.9	65.2	65.0	15.2	0.9	14.3	23.4	5.7	23.9
1974	59.0	64.2	63.3	15.7	1.5	14.2	24.8	9.5	24.0
1975	51.0	54.6	54.1	13.1	1.1	12.0	24.2	8.3	23.5
1976	58.3	62.6	62.0	15.4	1.6	13.8	24.9	10.4	23.7
1977	60.0	64.8	62.2	16.3	2.1	14.3	26.3	12.6	23.8
1978	62.0	68.1	67.7	16.8	1.8	15.0	24.8	10.8	24.1
1979	64.3	70.3	69.8	18.0	2.5	15.5	25.8	13.9	24.1
1980	63.4	67.8	67.2	18.0	2.9	15.1	26.7	16.0	23.8
1981	64.3	68.7	68.0	17.7	2.4	15.2	26.0	13.8	23.7

1982	61.0	65.1	64.7	17.1	2.5	14.6	26.4	14.4	24.0
1983	66.8	71.7	71.2	18.7	2.9	15.8	26.3	15.5	23.7
1984	70.3	77.3	76.9	20.5	3.6	16.9	26.7	17.8	24.0
1985	68.7	76.2	76.1	20.4	3.8	16.6	26.8	18.7	24.1
1986	72.5	79.7	79.8	22.5	4.4	18.1	28.2	19.5	25.0
1987	75.9	83.9	83.5	24.0	5.1	18.9	28.8	21.4	24.9
1988	78.0	86.1	85.5	26.2	6.3	19.9	30.6	24.0	25.5
1989	78.4	85.8	85.2	27.1	6.2	20.9	31.8	22.9	26.7
1990	80.3	87.6	86.7	29.1	6.4	22.7	33.6	22.0	28.3
1991	81.1	86.0	84.9	31.2	6.5	24.7	36.7	20.8	30.5
1992	84.6	89.5	88.1	33.6	6.3	27.3	38.1	18.8	32.3
1993	86.6	92.6	91.5	34.9	5.9	28.0	38.2	16.9	32,3
1994	89.6	94.6	93.6	35.8	6.8	29.2	38.2	19.1	32.5
2000	100.0	103.8	102.9	51.0	7.7	39.7	49.6	15.2	39.7
			High export scenario:		9.6	37.8		18.8	37.8

*In domestic mills, does not include other uses estimated to be 3.6 million tons in 2000.
Sources: Franklin Associates, Ltd., *Paper Recycling: The View to 1995* (1990) and *The Outlook for Paper Recovery to the Year 2000* (1993) (for 1970–92 and 2000); *Lockwood-Post's Directory* (annual).

Information on recovery rates prior to 1970 is somewhat more difficult to come by. In earlier decades, wastepaper was viewed from perspectives other than that of waste management—primarily in terms of national timber security; thus, historical figures report wastepaper use, but not wastepaper recovery. After the year 1970, however, the export of domestic wastepaper clearly became significant. By that year the wastepaper export rate (percent of wastepaper recovered) was about only 4 percent. As the recovery rate gradually increased, however, the export rate rose even faster, climbing to well above 20 percent in the late 1980s and early 1990s (in California the out-of-state wastepaper export rate was estimated at about 55 percent in 1990).[15] The AFPA projects continued substantial exports (15–20 percent) to the year 2000. Nearly all the growth in wastepaper recovery between 1970 and the mid-1980s was driven by export demand.

When large proportions of recovered wastepaper are exported, of course, the solid waste diversion benefits are still captured, but the anticipated domestic energy, fiber, and pollution prevention values of wastepaper begin to disappear. The wastepaper utilization rate provides an alternative indication of whether these advantages, ultimately tied to the effect of increased wastepaper use on domestic woodpulp production, are also being gained. As of 1994, about nine years since the latest upward trend in the wastepaper recovery rate appeared, the most accurate answer is "partially, at best." To understand this qualification, it is first helpful to consider the history of fiber utilization by the paper industry, as depicted in figure 4.1 and table 4.2 (see the detailed discussion of sources and assumptions in the notes).[16]

For the first half of this century, wastepaper accounted for a third or more of all fiber sources used in paper production. The highest levels of wastepaper use were seen during World War II and immediately after, when the wastepaper utilization rate (UR) reached nearly 40 percent. By the late 1950s, however, wastepaper use had begun to decline, falling below 20 percent in the 1960s and hovering in that range until the mid-1980s. Around 1985–1986, reflecting the increase in the wastepaper recovery rate, wastepaper utilization itself began to rise, and the wastepaper UR climbed more than 8 points, from about 24 to 32 percent between 1985 and 1993. Thus, although we have indeed gained

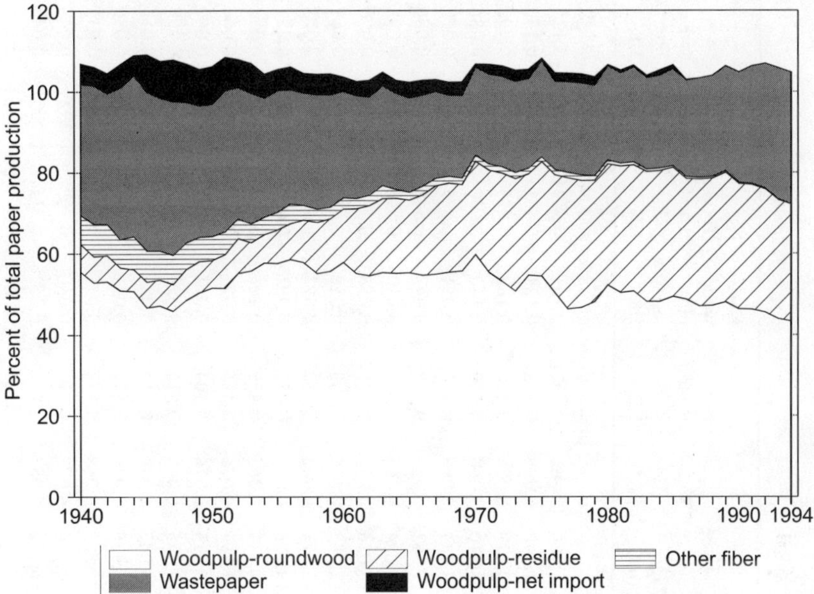

Figure 4.1
Fiber utilization, 1940–1994
Source: Table 4.2.

significant ground in recent years, it was not until about 1993 that the industry began to approach the levels of wastepaper utilization characteristic of the first half of the century. It is not expected to approach the wastepaper UR characteristic of the 1940s (near 40 percent) until the next century, and it lags far behind other major producing countries such as Germany and Japan with wastepaper URs already at around 53 percent. From an historical context the recent gains, although significant, are somewhat disproportionate to the hyperbole with which they have been greeted: "What's occurring in the paper industry, particularly with respect to wastepaper, is nothing short of revolutionary. The entire structure of the industry has changed."[17]

The picture is more complex for primary and secondary virgin wood use. As wastepaper use began to fall in the late 1950s, the use of wood residues primarily from sawmills began to rise rapidly, with residue-based domestic woodpulp utilization gaining almost 10 percentage points in

Table 4.2
Fiber utilization in paper production, 1919–1994

	Wood-pulp prod.	Wood-pulp exports	Wood-pulp imports	Net wood-pulp imports	Actual wood-pulp cons.	R^*	Waste-paper cons.	Other fiber cons.	Paper prod.	Wood residue UR*	Round-wood UR	Net wood-pulp imports UR	Total wood-pulp UR	Waste-paper UR	Other fiber UR	Total fiber UR
					(1,000 tons)					*(tons fiber used per ton paper and paperboard produced)*						
1919	na	na	na	na	4,020	na	1,854	748	5,966	na	na	na	0.674	0.311	0.125	1.110
1929	4,863	54	1,881	1,827	6,289	0.25	3,842	1,443	11,140	0.099	0.302	0.164	0.565	0.345	0.130	1.039
1935	4,926	172	1,933	1,761	6,442	0.17	3,587	969	10,479	0.076	0.370	0.168	0.615	0.342	0.092	1.050
1939	6,993	140	2,026	1,886	8,650	0.12	4,366	1,161	13,510	0.060	0.440	0.140	0.640	0.323	0.086	1.049
1940	8,960	481	1,225	744	9,782	0.11	4,668	1,044	14,484	0.069	0.555	0.051	0.675	0.322	0.072	1.070
1941	10,375	329	1,158	829	11,364	0.11	6,075	1,418	17,762	0.064	0.529	0.047	0.640	0.342	0.080	1.062
1942	10,783	378	1,237	859	11,038	0.11	5,495	1,325	17,084	0.065	0.531	0.050	0.646	0.322	0.078	1.045
1943	9,680	301	1,306	1,005	10,635	0.10	6,368	1,196	17,036	0.057	0.509	0.059	0.624	0.374	0.070	1.068
1944	10,108	218	1,072	854	10,502	0.10	6,859	1,385	17,183	0.054	0.508	0.050	0.611	0.399	0.081	1.091
1945	10,167	135	1,754	1,619	10,825	0.12	6,800	1,344	17,371	0.063	0.467	0.093	0.623	0.391	0.077	1.092
1946	10,607	39	1,805	1,766	12,092	0.12	7,278	1,382	19,278	0.063	0.472	0.092	0.627	0.378	0.072	1.076
1947	11,946	130	2,322	2,192	13,253	0.12	8,009	1,526	21,114	0.065	0.459	0.104	0.628	0.379	0.072	1.079
1948	12,872	94	2,176	2,082	14,375	0.14	7,585	1,452	21,897	0.076	0.485	0.095	0.656	0.346	0.066	1.069
1949	12,207	122	1,963	1,841	13,636	0.13	6,600	1,215	20,315	0.076	0.504	0.091	0.671	0.325	0.060	1.056
1950	14,849	96	2,385	2,289	16,509	0.12	7,956	1,439	24,375	0.068	0.515	0.094	0.677	0.326	0.059	1.063
1951	16,524	202	2,361	2,159	17,737	0.14	9,071	1,457	26,047	0.084	0.514	0.083	0.681	0.348	0.056	1.085
1952	16,473	212	1,937	1,725	17,286	0.14	7,881	1,211	24,418	0.086	0.551	0.071	0.708	0.323	0.050	1.080
1953	17,537	162	2,158	1,996	18,684	0.11	8,531	1,255	26,605	0.069	0.558	0.075	0.702	0.321	0.047	1.070
1954	18,256	442	2,051	1,609	18,989	0.11	7,857	1,200	26,876	0.070	0.577	0.060	0.707	0.292	0.045	1.044
1955	20,740	631	2,214	1,583	21,454	0.12	9,041	1,340	30,178	0.082	0.577	0.052	0.711	0.300	0.044	1.055

1956	22,131	525	2,332	1,807	22,998	0.13	8,836	1,551	31,441	0.088	0.586	0.057	0.731	0.281	0.049	1.062
1957	21,800	622	2,101	1,479	22,459	0.16	8,493	1,105	30,666	0.106	0.578	0.048	0.732	0.277	0.036	1.045
1958	21,796	515	2,105	1,590	22,483	0.19	8,671	1,003	30,823	0.127	0.550	0.052	0.729	0.281	0.033	1.043
1959	24,383	653	2,431	1,778	25,155	0.19	9,414	979	34,036	0.131	0.556	0.052	0.739	0.277	0.029	1.044
1960	25,316	1,142	2,389	1,247	25,700	0.19	9,032	971	34,444	0.132	0.578	0.036	0.746	0.262	0.028	1.037
1961	26,253	1,178	2,467	1,289	26,683	0.23	9,018	894	35,698	0.160	0.551	0.036	0.747	0.253	0.025	1.025
1962	27,908	1,186	2,789	1,603	28,598	0.24	9,075	963	37,543	0.174	0.545	0.043	0.762	0.242	0.026	1.029
1963	30,121	1,422	2,775	1,353	30,220	0.25	9,613	1,285	39,231	0.182	0.553	0.034	0.770	0.245	0.033	1.048
1964	32,415	1,580	2,942	1,362	32,088	0.25	9,843	929	41,703	0.185	0.552	0.033	0.769	0.236	0.022	1.028
1965	33,993	1,402	3,130	1,728	34,006	0.25	10,231	879	44,091	0.180	0.552	0.039	0.771	0.232	0.020	1.023
1966	36,603	1,572	3,357	1,785	36,922	0.27	10,564	980	47,113	0.200	0.546	0.038	0.784	0.224	0.021	1.029
1967	36,677	1,721	3,170	1,449	36,994	0.29	9,888	836	46,296	0.219	0.549	0.031	0.799	0.214	0.018	1.031
1968	40,892	1,902	3,532	1,630	41,303	0.28	10,222	905	51,245	0.220	0.554	0.032	0.806	0.199	0.018	1.023
1969	42,813	2,103	4,040	1,937	43,700	0.28	10,939	878	54,187	0.214	0.557	0.036	0.806	0.202	0.016	1.025
1970	43,546	3,095	3,518	423	43,192	0.28	11,199	828	51,671	0.230	0.597	0.008	0.836	0.217	0.016	1.069
1971	43,903	2,175	3,515	1,340	44,148	0.31	11,758	875	53,163	0.252	0.553	0.025	0.830	0.221	0.016	1.068
1972	46,767	2,252	3,728	1,476	47,347	0.33	12,818	892	57,434	0.267	0.531	0.026	0.824	0.223	0.016	1.063
1973	48,327	2,344	4,002	1,658	48,722	0.35	13,456	883	59,900	0.279	0.507	0.028	0.813	0.225	0.015	1.053
1974	48,349	2,802	4,123	1,321	48,341	0.31	13,105	838	59,040	0.251	0.546	0.022	0.819	0.222	0.014	1.055
1975	43,084	2,782	3,078	296	42,431	0.34	12,115	625	50,976	0.282	0.545	0.006	0.832	0.238	0.012	1.082
1976	47,721	2,518	3,727	1,209	47,541	0.37	12,742	742	58,329	0.292	0.503	0.021	0.815	0.218	0.013	1.046
1977	49,132	2,640	3,871	1,231	48,477	0.41	13,455	826	60,040	0.324	0.463	0.021	0.807	0.224	0.014	1.045
1978	50,020	2,599	4,023	1,424	49,834	0.40	14,059	854	62,047	0.313	0.467	0.023	0.803	0.227	0.014	1.044
1979	50,612	2,935	4,318	1,383	51,623	0.38	14,367	727	64,345	0.300	0.481	0.021	0.802	0.223	0.011	1.037
1980	52,958	3,806	4,051	245	52,448	0.36	14,737	602	63,536	0.300	0.522	0.004	0.825	0.232	0.009	1.067
1981	52,790	3,678	4,087	409	52,779	0.38	14,374	510	64,259	0.311	0.504	0.006	0.821	0.224	0.008	1.053
1982	50,986	3,395	3,656	261	50,187	0.38	14,317	396	60,962	0.312	0.507	0.004	0.823	0.235	0.006	1.065
1983	54,055	3,644	4,093	449	53,970	0.40	15,087	413	66,764	0.320	0.482	0.007	0.808	0.226	0.006	1.041
1984	56,120	3,594	4,490	896	57,466	0.40	16,085	438	70,286	0.323	0.482	0.013	0.818	0.229	0.006	1.053

Table 4.2 (continued)

	Wood-pulp prod.	Wood-pulp exports	Wood-pulp imports	Net wood-pulp imports	Actual wood-pulp cons.	*R**	Waste-paper cons.	Other fiber cons.	Paper prod.	Wood residue UR*	Round-wood UR	Net wood-pulp imports UR	Total wood-pulp UR	Waste-paper UR	Other fiber UR	Total fiber UR
1985	54,679	3,791	4,466	675	56,639	0.39	16,431	301	68,829	0.319	0.494	0.010	0.823	0.239	0.004	1.066
1986	57,552	4,458	4,582	124	57061	0.38	17,276	313	72,490	0.298	0.487	0.002	0.787	0.238	0.004	1.030
1987	60,102	4,889	4,850	(39)	59474	0.4	18,627	319	75,918	0.314	0.470	0.000	0.783	0.245	0.004	1.033
1988	61,711	5,528	5,038	(490)	60847	0.4	19,457	319	78,003	0.315	0.472	0.000	0.780	0.249	0.004	1.034
1989	62,598	6,231	5,005	(1,226)	61500	0.4	20,277	318	78,354	0.320	0.480	0.000	0.785	0.259	0.004	1.048
1990	63,667	5,905	4,893	(1,012)	61015	0.4	21,876	268	80,283	0.309	0.464	0.000	0.760	0.272	0.003	1.036
1991	64,418	6,338	4,997	(1,341)	61153	0.4	23,518	187	81,100	0.308	0.462	0.000	0.754	0.290	0.002	1.046
1992	65,942	7,222	5,029	(2,193)	62011	0.4	26,016	177	84,559	0.304	0.456	0.000	0.733	0.308	0.002	1.043
1993	63,507	6,499	5,413	(1,086)	62356	0.4	27,995	187	86,558	0.293	0.440	0.000	0.720	0.323	0.002	1.046
1994	64,600	6,662	5,637	(1,025)	63629	0.4	29,158	187	89,580	0.289	0.433	0.000	0.710	0.325	0.000	1.036

*R = estimated ratio of wood residue to primary pulpwood used. UR = utilization rate (tons fiber used per ton paper and paperboard produced).

Sources: U.S. Dept. of Agriculture, Forest Service (1982, 1990); U.S. Dept. of Commerce (1992); Franklin Associates (1990, 1993); *Lockwood Post's Directory* (annual); FAO (1991); plus assumptions and calculations. See detailed discussion of sources and assumptions in notes.

each of the next two decades. Prior to the late 1950s wood residue had accounted for less than 10 percent of paper production and for half that during the war years of the 1940s, but by 1980 it had risen into the 30 percent range, where it has remained. As the use of wood residue rose and the use of wastepaper fell relative to total production, the use of roundwood (logs) also fell, declining from a roundwood UR of 50–60 percent to roughly 45 percent in recent years.

Two other factors, although of limited importance in recent decades, must also be considered. In the 1920s and 1930s, both woodpulp imports and nonwood fiber sources ("other fiber") accounted for significant proportions of the materials used in paper production. The UR of nonwood fiber sources (such as rags, cotton linters, and wheat straw), for example, was up to 13 percent in the second and third decades of the century, but has declined steadily to less than half of one percent at present.

Net woodpulp imports provided almost 17 percent of all fiber sources used in paper production in the mid-1930s, representing almost a quarter of all woodpulp consumed. However, woodpulp imports were sharply curtailed during World War II, declining by 65 percent between 1939 and 1940. This change was likely a factor in the high wastepaper utilization rates of the period. After the war, imports picked up again, but fell off steadily in importance in subsequent decades, with a net woodpulp import UR declining from about 5–6 percent in the mid-1950s to about 1 percent in the 1980s. In 1987, apparently reversing more than a century of history, precisely at the same time as the wastepaper UR increase, the balance of trade in woodpulp shifted, and the United States became, for the first time, a net exporter of woodpulp. By the early 1990s net woodpulp exports represented about 3 percent of woodpulp production (actual exports were about 10 percent of production).

Overall, the relative decline in the use of logs, wastepaper, nonwood fibers and imported woodpulp over the last five decades has more than been made up for by the increased use of wood residue—essentially amounting to a simplification in the fiber resource base for production, and most likely not unrelated to the increasing scale and rigidity of production formats. The net result is that the total proportion of virgin woodpulp from all sources (roundwood, residue, net woodpulp imports) increased steadily from about 63 percent of production at the end of

World War II, to 75–85 percent, where it stayed for the past three decades, just beginning to show a decline around 1990.

The recently increased proportion of wastepaper in domestic paper and paperboard appears to have influenced woodpulp *consumption* more strongly than woodpulp *production*. In the nine years between 1985 and 1994, wastepaper use grew rapidly: 6.5 percent per year, or about triple its rate of growth in the preceding decade. (All rates are compounded.) In the same period, annual growth in paper production, woodpulp production, and woodpulp consumption was, respectively, 3.0, 1.9, and 1.3 percent. Only the rate of growth in woodpulp consumption declined significantly (by more than half) from its rate of growth in the preceding decade (1975–85), whereas woodpulp production grew about 20 percent more slowly and paper production grew at roughly the same rate. In the same period, however, annual growth in woodpulp exports was 6.5 percent, or triple the rate of growth in the preceding decade (growth in woodpulp imports averaged 2.6 percent, slower than in the preceding decade).

Thus, although the comparatively rapid growth in wastepaper utilization appears to have displaced some woodpulp consumption that would otherwise have been expected, the corresponding effect on woodpulp production appears more limited because much of the differential was channeled into substantially increased woodpulp exports. A magnification of this effect—diluting the prospective wood use reduction benefits associated with wastepaper use through increased upstream fiber exports—seems to have also occurred with pulpwood. In the same nine-year period, gross exports of pulpwood chips more than doubled, growing at an annual rate of almost 11 percent (relative to negative growth in the preceding decade), and chip imports continued to decline.[18]

Another possible effect of the increased use of wastepaper is the degree to which it may constitute not only a supplement (rather than substitute) to total pulp production, but also to total paper production. Although annual growth in paper production in the nine-year period considered was consistent with growth in the preceding decade, it grew a third faster than domestic paper consumption in the latter period. This was reflected in the continuing substantial growth in paper exports, which was beginning to close the longstanding gap in net trade of paper commodities.

Whereas paper imports grew at about 2.4 percent per year in the nine-year period (a third the rate of the preceding decade), paper exports grew at almost 11 percent per year, two and a half times the rate of growth in the preceding decade.

Although these export-oriented developments may be relatively short-term adjustments to a wastepaper utilization rate that accelerated in response to the recycling movement of the 1980s, they highlight the significance of shifts in the upstream and downstream forks of the wood flow in terms of translating recycling into forest conservation (reduced wood use) or pollution prevention (reduced woodpulp production).

Does paper recycling save trees? The simplest answer is "not by itself," or at least, "not yet." Others have stated, more bluntly, that various related adjustments in timber products demand and allocation "cast serious doubt on the view that increased wastepaper recycling will save significant amounts of valuable forestlands."[19] Increased wastepaper utilization clearly creates the potential for reduced wood use (or at least reduced growth in wood use) by the domestic industry, and thus for reduced wood use overall, but nothing in either recovery or utilization rates tells us enough to know how or even if this potential translates back to the resource level. In looking upstream to find out, however, we leave the domain of solid waste and head back into the territory of forest resource allocation and trade. In this land, where the inhabitants speak an entirely different language, the question is not, "Does paper recycling save trees?" but rather, as put by a prominent forest economist, "Will increased paper recycling extend timber resources?"

In an important recent study, *Recycling and Long-Range Timber Outlook* (1994), USDA economist Peter Ince examined the relationship between wastepaper use and timber demand. The analysis was based on the North American Pulp and Paper (NAPAP) sector economic model, jointly developed by the USDA and Forestry Canada (the Canadian ministry of forests). Conducted in support of the 1993 RPA Assessment Update, it was the first major study that attempted to synthesize recent trends in recycling and long-range timber markets systematically. It was also an outcome of the 1989 RPA Assessment. Constructed with a more limited economic model, around data series that terminated at almost the precise time that wastepaper recovery and utilization took an upturn, the

1989 report had projected levels of wastepaper recovery and utilization in the year 2040 that were in fact reached by 1992!

The recent report based on the NAPAP model examined a wide range of trends and interactions for the period 1990–2040. In this analysis, paper production was projected to rise about 70 percent during the period, to around 135 million tpy. Wastepaper recovery was projected to increase to 49 percent by 2000 and to 60 percent by 2040. Due to rising consumption this would not actually decrease the gross annual wastepaper disposal burden, but it would stop its growth. Wastepaper utilization was projected to grow more slowly than it has in recent years, eventually coming close to the 50 percent level (about 48 percent) by 2040—lagging a half-century or so behind Germany, Japan and other major producing countries. The timber harvest was projected to increase by about 40 percent. In summarizing the analysis, Peter Ince concluded:

> With more abundant fiber resources (due to increased recycling) paper and paperboard consumption, . . . and production of pulp, paper and paperboard commodities were projected to increase very substantially in the decades ahead. Exports of paper and paperboard products were also projected to grow substantially, while imports were projected to generally decline. . . . Thus, although recycling rates were projected to increase, total consumption and export of pulpwood (pulpwood supply) was projected to increase.[20]

For the record, despite an expected 70 percent increase in production, total employment in the industry was expected to decrease.

The good news is that this is a forecast, not a sentence. As Ince reminds us, the "potential for change in government policies is an important part of the issue." Indeed, the wildly inaccurate recycling scenarios forecast in the 1989 assessment in part reflected an inability to predict changes arising from shifts in public opinion and government policy. Ince and others have elsewhere examined the significant potential of various interventions to change the current projections—including increased consumer education on reducing paper use, increased landfill tipping fees, the imposition of advance waste disposal fees on producers, and investment tax credits for expanded recovered materials utilization capacity.[21] Needless to say, recent and prospective interventions at the timber resource level and by the EPA are also significant factors. Lately, efforts have been proposed to incorporate various nonwood fiber-use scenarios into the NAPAP model, suggesting further interactions and influences that may alter the long-range trends currently predicted.[22]

The bad news is—given current policies and trends in resource management, production, consumption, exports, and waste management—the present correlation between paper recycling and *wood-use reduction* is extremely weak (although the analysis does find that the ability to support expected future demand has been extended). This more complex analysis may be deeply troubling to those who have grown accustomed to the assumption that "each ton of recycled paper saves seventeen trees" or that there is a one-for-one correspondence between recovered materials and virgin materials use. Indeed, some recycling advocates have suggested that such analyses should be downplayed lest they dampen public support for recycling and endanger the prospects for maintaining still-needed pressure to increase wastepaper recovery and utilization. Others have admitted to some culpability for having oversold (or at least oversimplified) the effects of recycling, and thus having undersold the need for effort in related areas. This more thoughtful approach is a step in the right direction although it underscores the tensions between the need to mobilize, and the need to educate.[23]

Obviously, this is not an argument for abandoning the hope that wastepaper use will have a significant role to play in wood-use reduction, but it does call for greater recognition that the apparent success of recent years, measured by recovery and utilization rates, is heavily tempered by related shifts in other factors. The historical context points out the loss of diversity in fiber resources and the degree of slippage in wastepaper use that occurred in the past half-century. The modern international context indicates that the domestic industry has a long way to go to catch up with the achievements of other countries and illustrates the degree to which exports can siphon off apparent gains. The inexorable growth in production and consumption continues to overshadow everything.

Recovered Fiber Grades

In practice, hundreds of categories defined by precise technical criteria (basis weight, brightness, strength, etc.) are used in the trade and marketing of paper and paperboard commodities; dozens are used in the trade of wastepaper. For statistical and evaluation purposes a more limited set of paper and wastepaper grades than those actually used in trade are commonly monitored. RCRA, as elaborated by the EPA, has provided its own set of terms in guidelines for federal procurement of items containing the highest practicable percentage of recovered materials. The various terms pertinent to wastepaper are outlined below.[24]

Recovered fiber refers to all recovered materials suitable for use in paper production—theoretically, the actually recovered portion of all fibers that become available as by-products of production and use. This may include logging and sawmill residue as well as urban wood waste, cellulosic textile and cordage wastes, agricultural waste, and all forms of wastepaper. In the RCRA context, however, recovered fiber refers only to materials recovered from RCRA-defined solid waste, which (based on the municipal/industrial differentiation) excludes wastes generated in the original paper production process, as well as forest residues, although it does not exclude recovered industrial by-products of textile, agricultural or other production activities. The exclusion is largely focused on *mill broke,* or wet fiber residue, trimmings, and rejects from the paper machine and main paper reel (i.e., materials generated before the papermaking process is complete). More recently, the definition of mill broke has been expanded to include *other mill waste* such as butt rolls and obsolete or off-specification inventory. (Ideally, however, anything included within gross paper production figures, even if never shipped, should be included within the definition of recovered materials to preserve materials balance correlations). *Wastepaper* was viewed in the original procurement guidelines as the subset of recovered paper materials (now recovered fiber) that are actually paper. The term wastepaper has since been dropped, and wastepaper is simply included within the definition of recovered fiber.

Pulp substitutes are an unprinted postmill/preconsumer grade of wastepaper that includes trimmings, mill wrappers, obsolete stock, butt rolls, and related scrap resulting from the conversion of bulk paper and paperboard stock into end-consumer products such as boxes, stationary, forms, and publications. In practice, corrugated container and recycled boxboard converter scraps, along with unprinted newsprint, are tracked within other wastepaper grades and not within pulp substitutes, which, as tracked, are predominately white paper(board) grades, with high chemical woodpulp content.

Postconsumer fiber (formerly *postconsumer paper materials*) refers to materials that have passed through their end-usage as a consumer item, and include old newspapers (ONP), old magazines (OMG), used corrugated containers (OCC), office wastepaper (OWP), and miscellaneous (mixed) paper and boxboard. The term *end-usage* has been clarified to refer to the last intended consumer, rather than the last actual consumer. Postconsumer fiber therefore excludes pulp substitutes, along with the misprints and overruns from commercial printers and publishers that, together with clean white OWP, constitute much of the wastepaper grade known as high-grade deinking. The RCRA amendments of 1984 called for maximizing the use of postconsumer recovered materials, but they also retained an interest in encouraging the use of preconsumer recovered materials. The latter had long been recovered and used at high rates and thus were not a significant factor in the solid waste crisis that drove the focus on postconsumer material recovery and utilization.

The primary categories commonly used to track the recovery—and thus trade and use—of wastepaper are *pulp substitutes, newsprint, corrugated, high-grade deinking,* and *mixed.* There is no exact association between these trade terms, and the postconsumer/preconsumer distinction (except in the case of pulp substitutes); mixed papers tend to be postconsumer, whereas newsprint, corrugated and deinking grades include both pre- and postconsumer material. One cannot necessarily distinguish a postconsumer paper from a preconsumer paper by looking at it, nor by listing its material properties. Thus, a chain of record keeping must be instituted (somewhat akin to the manifest system instituted for RCRA-defined hazardous waste) stretching from waste haulers to paper brokers to mills and beyond if mills are to be able to validate the postconsumer fiber content of the products they manufacture and if procurement officers are to validate that they have met their obligation to purchase or use recycled paper products as defined under RCRA and various state and local mandates.

Figure 4.2 outlines the flow of paper and wastepaper from production through consumption and recovery for 1992.[25] In that year, 84.6 million tons of paper(board) was produced, and net imports added about 3.6 million tons, for a total "new supply" (total apparent consumption) of about 88 million tons. Of this, 38 percent was recovered, and 62 percent was disposed. Corrugated (45 percent) and newsprint (21 percent) grades

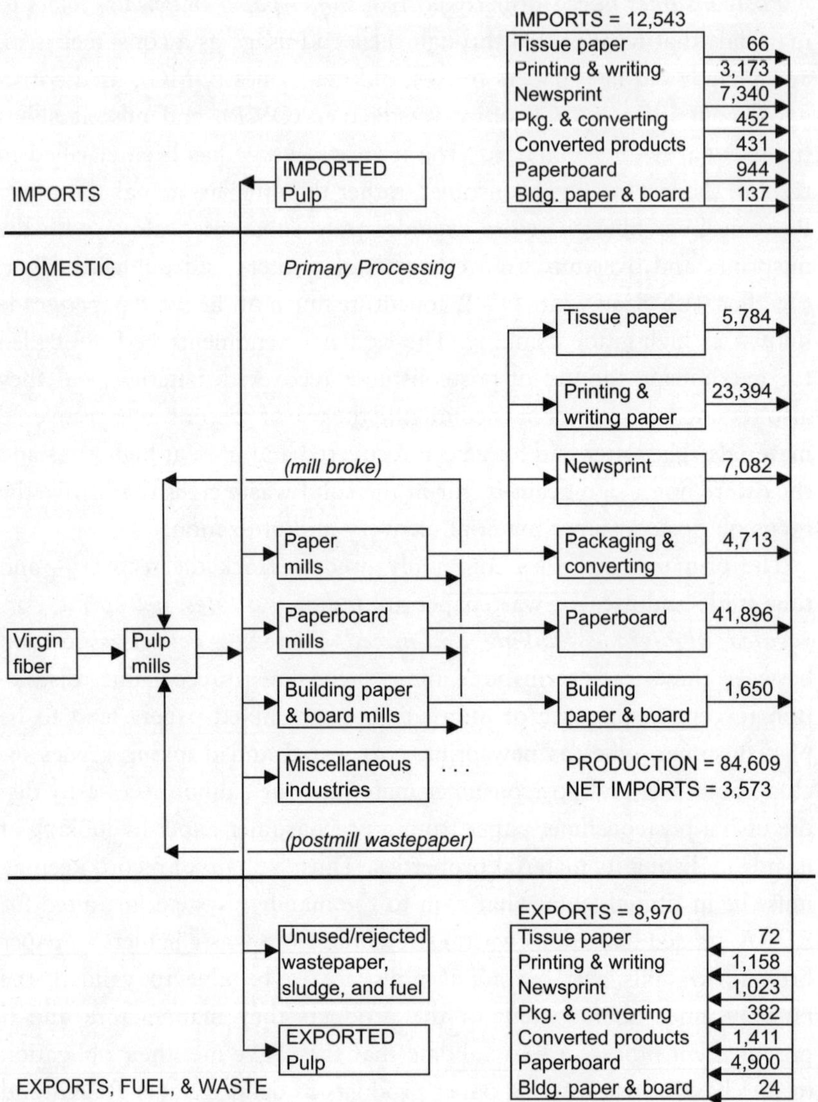

Figure 4.2
Paper flow, 1992
Sources: *Lockwood-Post's Directory* (annual) (for production); Franklin Associates, Ltd., *The Outlook for Paper Recovery to the Year 2000* (1993) (for wastepaper).

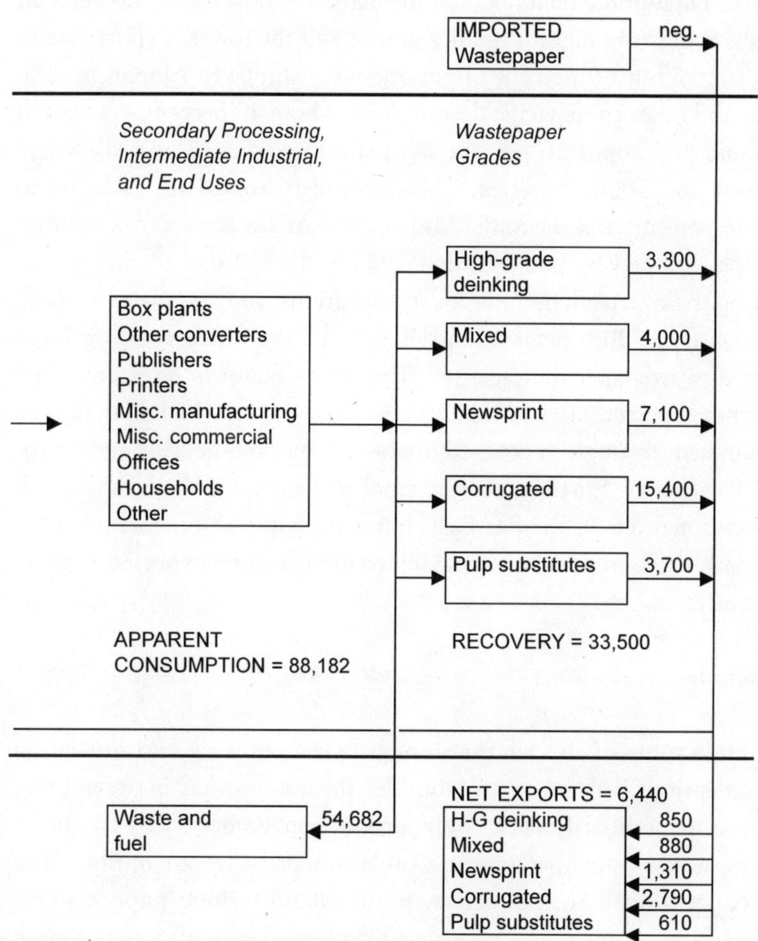

IMPORTED
Wastepaper
neg.
Secondary Processing,
Intermediate Industrial,
and End Uses
Wastepaper
Grades
High-grade
deinking
3,300
Box plants
Other converters
Publishers
Printers
Misc. manufacturing
Misc. commercial
Offices
Households
Other
Mixed
4,000
Newsprint
7,100
Corrugated
15,400
Pulp substitutes
3,700
APPARENT
CONSUMPTION = 88,182
RECOVERY = 33,500
NET EXPORTS = 6,440
Waste and
fuel
54,682
H-G deinking
850
Mixed
880
Newsprint
1,310
Corrugated
2,790
Pulp substitutes
610

accounted for two-thirds of recovery, followed by pulp substitutes (12 percent), mixed (12 percent), and high-grade deinking (10 percent).

Net wastepaper exports consumed about 20 percent of all wastepaper recovered. High-grade deinking had the highest export rate at 26 percent; pulp substitutes (the highest-quality grade) had the lowest export rate at 17 percent. About 60 percent of exports were shipped to Japan, Korea, Taiwan, and other transpacific destinations. About 12 percent went north to Canada, and about 20 percent went south to Mexico and elsewhere in Central and South America. The remainder crossed the Atlantic to Europe (6 percent) and the to the Mideast and Africa (2 percent). Exports to Europe dropped by half between 1989 and 1992 due to the institution of aggressive packaging recovery programs and mandates in Germany and other European countries and to the corresponding large supplies of wastepaper that accrued. This was a factor in dampening the even higher expected rates of U.S. wastepaper exports, which had, in fact, been pursued through recovered materials market development programs. Exports of European wastepaper are expected to dampen U.S. wastepaper markets in the Far East, but wastepaper shipments from the United States to both Canada and Mexico are presently expected to grow significantly.

Paper Grades

To examine some of the relationships between recovery and utilization in more depth, it is necessary to consider the associations between pulp, paper and paperboard stock (bulk production grades), end-consumer products, and wastepaper grades. Table 4.3 provides an outline. The major paper grades are newsprint, printing and writing papers, tissue papers, and packaging and converting papers. The major paperboard grades are defined by their primary pulp basis: bleached kraft, unbleached kraft, semichemical, and mixed furnish or recycled paperboard.[26] Building paper and paperboard, including products such as asphalt roofing paper, constitute a third production class—a comparatively inconsequential class as substitute building products based on synthetic materials have made increasing inroads. Total domestic production is divided evenly between paper (48 percent) and paperboard (49 percent), with building grades accounting for less than 2 percent. (All figures in this section are from 1992.)

Almost all *unbleached kraft paperboard,* which represents about a quarter of total production, is produced for use as linerboard, the facing material in corrugated boxes. Essentially all semichemical pulp is used in *semichemical paperboard,* and almost all of this is used in the fluted corrugating medium sandwiched between linerboard facings. Corrugating medium also represents a significant application for recycled paperboard. Altogether, corrugated boxes, also known as containerboard, account for over a third of total production in the paper industry.

Recycled paperboard is predominantly used in folding boxboard as packaging for an almost endless array of consumer products (cereal boxes, detergent boxes, etc.), as well as in corrugating medium, linerboard, and a variety of tubes, cans, and drums. *Bleached kraft paperboard* is primarily used for food service containers (like milk cartons or pizza boxes), and in other boxboard or linerboard applications.

Printing and writing papers (which by convention do not include newsprint) are the fastest-growing segment of the industry and presently represent slightly less than 30 percent of total production. Groundwood papers, often coated with clay to improve printing qualities, are used in magazines, advertising inserts, catalogues, phone directories, pulp novels (lower-grade paperback books) and the like. They are sometimes called "wood-containing paper," generally referring to the high lignin content deriving from the use of mechanical pulp. Free-sheet papers ("wood-free") are based on bleached chemical pulp. Coated applications include glossy catalogues and higher-grade magazine papers. The dominant uncoated free-sheet category, about 15 percent of total production, includes most office papers (copier, forms, envelopes). Specialty categories include bleached bristols, stiff card papers used for covers and folders, thin papers such as those used in encyclopedias, and cotton paper used in premium quality free-sheet applications. (U.S. paper currency is based on a cotton/linen blend.)

The other primary paper grades include *newsprint, tissue paper,* and *packaging and industrial converting paper.* Each accounts for roughly 8 percent of total production, although production of the latter has been

Table 4.3
Paper production, wastepaper use and recovery by grade, 1992

Wastepaper used (mil. tons)		Production by primary grade	Production (mil. tons)	Production (% total)	Annual growth (1982–92) (%)	Waste-paper utilization (% prod.)	Recovery (% cons.)
News	2.3	NEWSPRINT	7.082	8.4	3.5	46.3	56.3
Mixed	0.9						
PS	1.5	PRINTING & WRITING	23.394	27.7	4.1	9.4	29.2
Deink.	0.6	Groundwood	1.609	1.9			
Mixed	0.1	Coated	8.119	9.6			
		Uncoated free sheet	12.170	14.4			
		Bleached bristols	1.162	1.4			
		Cotton	0.159	0.2			
		Thin	0.175	0.2			
Deink.	1.7	TISSUE	5.784	6.8	2.7	60.5	8.62
PS	0.7	Toilet	2.382	2.8			
Mixed	0.5	Toweling	1.883	2.2			
News	0.5	Facial, napkin, other	1.519	1.8			
Corrug.	0.1						
Corrug.	0.2	PKG & CONVERTING	4.713	5.6	(1.2)	10.6	12.50
PS	0.2	Unbleached kraft	2.380	2.8			
Mixed	0.1	Bleached & speciality	2.333	2.8			
All	9.5	TOTAL PAPER	40.973	48.5	3.0	23.1	32.1

	0.0	SOLID BLEACHED PAPERBOARD milk carton, food service linerboard, boxboard	4.503	5.3	2.1	0.0	negligible, pulp subs. only
Corrug. Mixed Deink. PS News	3.2 0.1 0.1 0.1 0.1	UNBL. KRAFT PAPERBOARD linerboard, boxboard	21.658	25.6	4.1	15.5	see container- board
Corrug. Mixed News	2.1 0.1 0.1	SEMICHEMICAL PAPERBOARD corrugating medium	5.762	6.8	2.8	37.3	see container- board
Corrug. Mixed News PS Deink.	6.7 1.8 1.5 0.6 0.1	RECYCLED PAPERBOARD boxboard, containerboard, wallboard facing, tubes, cans, drums	10.063	11.9	4.5	105.9	see boxboard
		PAPERBOARD (by major application):					
All	*9.0*	*Containerboard*	*29.200*	*34.5*	*4.2*	*30.8*	*59.0*
		Boxboard					*18.4*
All	16.2	TOTAL PAPERBOARD	41.986	49.7	3.8	38.5	
All	0.8	BUILDING PAPER & BOARD	1.600	1.9	0.7	61.9	5.9
All	26.5	TOTAL PRODUCTION	84.559	100.0	3.3	31.3	38.0

Sources: Franklin Associates, Ltd., *The Outlook for Paper Recovery to the Year 2000* (1993) (wastepaper use); *Lockwood-Post's Directory* (annual) (production).

declining with the substitution of plastic for the brown paper bag (unbleached kraft). Newsprint is discussed at length in the final section of this chapter.

The highest wastepaper utilization rates are found in recycled paperboard (108 percent in 1992), tissue paper (60 percent), semichemical paperboard (38 percent) and newsprint (36 percent). Together, these products represent a third of production and 72 percent of total wastepaper utilization. The lowest wastepaper utilization rates have been in printing and writing papers (9 percent) and in bleached kraft paperboard because of restrictions historically associated with packaging that comes in contact with food.

The fastest increase in wastepaper utilization rates between 1982 and 1992 was in newsprint and tissue papers. Both had relatively low wastepaper utilization rates (22 and 39 percent) relative to their ability to accommodate up to 100 percent recycled content, and both could make use of the rapidly growing supplies of postconsumer wastepaper grades (ONP, OMG, mixed, and some deinking grades) without the more demanding (and expensive) deinking processes required for higher-end paper grades. However, of the almost 12 million ton increase in annual wastepaper use in the period, more than 60 percent was absorbed by paperboard grades.

In the next phase of growth, the emphasis, not surprisingly, will be on printing and writing papers. They are expected to continue to show the fastest rates of growth in both consumption and production. Associated wastepaper grades (increasingly better sorted mixed and deinking grades) will account for a growing proportion of recovered wastepaper. The majority of minimill projects recently announced or contemplated are focused on the production of deinked market pulp from office wastepaper.[27]

Paper Recycling and the Federal Government

A brief review of the history of RCRA and other federal recycling efforts and a survey of the more influential roles played by other factors at both federal and local levels provide important object lessons from recycling that may inform future, broader goals and interventions.

It is important to keep in mind that the preceding discussion has been based on aggregate classes. The thousands of products within these classes represent complex blends of pulps and various additives, each of which imparts different characteristics that must be balanced against not only wastepaper utilization interests, but also other environmental criteria and exacting performance standards. Relevant performance standards may include tear strength, burst strength, basis weight, stiffness, curl, folding characteristics, roughness, hue, brightness, opacity, porosity, gloss, ink-holding characteristics, dirt count, and permanence. One can easily understand that different pulp processes, based on different feedstocks, will yield pulp products with widely varied characteristics. Any of these characteristics may be mitigated or optimized, depending on the requirements of the intended application, through the blending of pulps; the use of mineral and synthetic fillers; various mechanical, thermal and chemical stock preparation processes; various paper machine processes including coating and calendering (a smoothing process); and converting processes.

One can equally well understand the aggravation of the consumer whose shopping bag rips on the way to the bus, the retailer whose produce arrives damaged, the commercial printer who loses a production run because the ink bled on the paper, the newspaper publisher who misses a deadline because the printing press keeps jamming, or the archivist who discovers the books are disintegrating. Papermaking has long been referred to as an art, and although much of the artistry may have been lost as production formats rigidified, it has been replaced by commercial standards that mediate these complex relationships between pulp, paper, and the rest of the economy.

Into this delicate picture, under court order to (essentially) quit stalling and follow the law, the EPA stepped in 1988 armed with a set of definitions that temporarily confused everyone and with its guidelines for the federal procurement of recycled paper. The term *recycled paper,* of course, presents particular difficulties. It is essentially shorthand for referring to paper (or paperboard) that has a "significant" level of recovered fiber content and, more specifically under the RCRA mandate, to paper that has the "maximum practicable" level of postconsumer content. Practicability, for different products, was to be determined with respect not only to technical performance standards, but also to standards of cost, competition, and availability.

The EPA had a very large number of variables to consider as it developed procurement guidelines, all of which were related to one another and related in various unique ways to evolving product-wastepaper-virgin-pulp configurations. The postconsumer content mandate not only required some form of manifest system, for example, but it also required the EPA to be involved at the level of matching particular wastepaper grades with particular paper products (by implication because particular wastepaper grades are predominately found in either pre- or postconsumer categories). If, indeed, the EPA guidelines were to set a standard that would be voluntarily adopted by a broader set of consumers and have an effect that rippled through the larger system of paper production and consumption, the product specifications would have to accommodate not only technical requirements and changing technologies, but also the relative availability of different wastepaper grades, the number of companies using them to produce various products, and the trends in each of these areas.

The EPA would further have to assure that standards would not introduce distortions that could misdirect wastepaper from its "best use." If, for example, aggressive standards for tissue paper or packaging products were accompanied by weak standards for other paper products, they could potentially cause the diversion of high-end wastepaper grades (e.g., high-grade deinking or pulp substitutes, with high levels of bleached chemical pulp content) away from high-end uses (e.g., uncoated printing and writing papers). These push/pull considerations were reflected in the EPA's comparatively flexible standards for printing and writing grade papers, wherein minimum content requirements were initially expressed in terms of *wastepaper* (now recovered fiber) rather than *postconsumer recovered materials* (now postconsumer fiber): "As more preconsumer waste paper is used for printing and writing papers, there will be less available as a raw material for other products. As a result, manufacturers will have to use more postconsumer recovered materials as a raw material."[28] This was, however, the most attacked of the product guidelines; many recycling advocates argued that the EPA should have set a postconsumer content standard for printing and writing grades. Ironically, in the late 1980s, due to the comparatively primitive level of development in domestic deinking technology, the use of postconsumer paper in this most demanding class of products (in comparison to tissue or newsprint) was extremely likely to be associated with chlorine-based bleaching of deinked pulps.[29]

By the late 1980s, essentially everyone, from Georgia-Pacific to Greenpeace, agreed that federal action was needed, at a minimum, to help rationalize competing definitions of recycled paper, and they had been waiting since 1981—the date by which RCRA had mandated that EPA publish procurement guidelines for paper. Environmentalists, concerned that the industry was increasingly just tacking a recycled label onto products incorporating the same old pulp substitutes and mill broke it had always used, wanted a more exacting definition so that markets could be created for the postconsumer wastepaper being collected (even the healthy export markets, which seemingly everyone viewed as an unconditional advantage, had some limitations) and so that would-be green consumers wouldn't be hoodwinked by misleading advertising. Furthermore, it was hoped the government would demonstrate that recycled papers of known composition really worked as well as virgin paper, thus putting old fears and rumors to rest. Paper manufacturers looked to the federal government to help eliminate the confusion in their markets as various state (even county and municipal) governments came up with a wide array of procurement policies and even wider-ranging mandates, such as the variously defined minimum recycled content standards for newsprint used by newspaper publishers in the state. Expectations were high and quite out of proportion to both the complexity of the task, and, ultimately, the limitations of the tool.

This tool—minimum content standards for federal procurement—in fact represented a dramatic reduction of the larger objectives that had been associated with municipal solid waste management and with recycling in particular for the preceding two decades. Although consuming a good deal of paper and symbolically important, the federal government nevertheless accounts for less than 3 percent of domestic paper consumption. Assuming (generously) that half of the government's paper purchases were affected by the limited standards that were finally published (and resoundingly criticized as both flimsy and incomplete), this still left about 99 percent or so of domestic paper consumption unaddressed in direct terms. However, if the content guidelines were widely followed by all levels of government, they could become more significant.[30]

When RCRA was passed in 1976, it contained two major components that addressed the recycling of MSW: (1) the development of federal procurement standards and supporting provisions, and (2) the provision of technical and financial assistance to the states.[31] Supporting provisions for the procurement guidelines required the Department of Commerce to develop uniform technical specifications for the classification of recovered materials and a "substitutability index" documenting the ability of recovered materials to replace virgin materials, and to identify markets and assess barriers to the use of recovered materials. The Office of Federal Procurement Policy (OFPP) (in the Executive Office of the President) was directed to coordinate the implementation of the RCRA procurement program with the EPA administrator and to prepare biannual reports to Congress on progress by federal agencies.

Although the Department of Commerce completed a few desultory studies, it never completed the substitutability index and, in any event, saw its funding for the project reduced by more than 75 percent during the 1980s. The OFPP issued two reports to Congress in which, as William Kovacs wrote, "The substance of the reports [revealed] the federal government's absolute failure to pursue either a policy of encouraging recycling or the procurement of items containing recovered materials." The first procurement guideline was issued five years late, and the paper guidelines were seven years late (and issued only after the EPA was sued by the Environmental Defense Fund). The key features of the state assistance provisions were financial assistance for state solid waste management planning efforts and the organization of Technical Assistance Panels. Appropriations for the former were reduced from $32 million to $0 between 1980 and 1984; the latter were specifically prohibited from being funded after 1981.[32]

The federal government's abandonment of recycling in the 1980s—in fact what looked more like its downright hostility to the idea when federal support of waste incineration is factored in—has been thoroughly documented, as have been the substantive efforts and progress made at the state level.[33] Ultimately, however, even if the approaches laid out by RCRA in the 1970s had been aggressively implemented, it seems uncertain whether they would have substantially changed the trajectory of developments in the wastepaper recovery rate, the wastepaper utilization rate, or the larger evolution of fiber utilization and trade patterns in the industry. The RCRA-style approaches to recycling, although at times framed within broader materials policy perspectives, were ultimately limited and weakened by a postconsumer waste focus. More than defining a federal role, they defined limits to the federal role—a role which, aside from the procurement policy (itself destined to be weakened by the complexity of writing mandatory content specifications), was substantially restricted to one of providing assistance.

RCRA also gave rise to a vocabulary for and cemented a view of recycling that is both fully inadequate with respect to the larger objectives recycling has been linked to and difficult to put back into a box. The RCRA postconsumer waste label functions at its best as an indicator of a larger problem in the production and consumption systems that in the case of wastepaper revolve around the paper industry, but it has developed a life of its own that far exceeds its ability to suggest successful approaches to the larger issues. Superficial effects include the puzzling, almost touchingly innocent conclusion reached by the publishers of one popular magazine in researching a possible switch from virgin to recycled paper stock in 1991: "We want to save landfill space and we want to save trees. The recycled paper now available, however, contains negligible amounts of postconsumer waste . . . effectively accomplishing neither goal. To use it, in our analysis, would be to cave into the hype without achieving any real progress."[34] More insidious effects of the RCRA isolation of postconsumer waste include the tendency to reinforce a preoccupation with products and consumers instead of with production. Products are important, and so are procurement decisions; however, to the extent that the accessibility and comparative friendliness of consumer-focused approaches robs the motivation to act in other, more difficult areas, they can be deeply misleading.

The procurement requirements for paper—the only survivor of the Reagan-era decimation of the RCRA approach to recycling—also had another side. When combined with the maximum feasible postconsumer mandate, the coverage of multiple, highly differentiated products, and the expectation that the guidelines might be widely adopted by other consumers, they potentially put the EPA into the business of mixing and matching particular products with particular wastepaper resources and of attempting to orchestrate complex forces of demand and allocation that would otherwise be left to the market. Market forces, of course, were largely working against increased wastepaper use both in terms of encouraging the concentration of the forest products industry and because deep-rooted failures in government policy around forest resources had left wood substantially undervalued. Intervention was definitely needed, and an industry-wide wastepaper utilization goal, in contrast to the wastepaper recovery goal that the industry set for the nation, could have gone a long way toward jump-starting increased recycling a decade or more before it was eventually stimulated by other forces. The EPA, however, was given a task that looked remarkably a lot like micromanagement of the marketplace—a task for which large government bureaucracies are arguably highly unsuited. That this unorthodox role for the EPA (with help from the Department of Commerce) was tolerated by industry, most likely represented the limited harmonizing effect it was expected to produce.

In the 1970s, the federal government had, in fact, attempted to establish wastepaper utilization goals for the industry. Under NECPA, the Department of Energy was required to develop, with industry participation, voluntary targets for industrial utilization of recovered materials. The measure was eventually repealed under intense fire from industry, although it provided an important historical precedent for directly associating multiple economic and environmental goals with the industrial utilization of recovered materials.[35] The opposition had been based on the argument that free market forces would provide the most effective stimuli for recycling. Although they might well provide the best allocation of wastepaper resources once activated, when left to the stimulus of market forces, the wastepaper utilization rate actually declined slightly after 1978 and did not pick up until mandatory materials recovery legislation was developed and passed by the states in the mid- to late 1980s. Ultimately, even mandatory procurement (consumer focused) was preferred over the most forgiving (producer focused) utilization goals.

The culmination of the 1980s' federal approach to recycling occurred in 1991. Three years earlier, operating once again under court order, the EPA had proposed a set of rules under the Clean Air Act for the regulation of municipal solid waste incinerators.[36] Among the rules was a relatively conservative requirement that 25 percent of the waste stream (by weight) in the incinerator service area be separated out for recycling. If no markets could be found for the materials, then the incinerator operator could burn the waste. This requirement was widely applauded by those who had been contesting the powerful industry forces promoting incinerators, which had, by virtue of flow-control ordinances granting rights to a given waste stream, conflicted with the ability of communities (and sometimes state requirements) to establish recycling programs. The EPA's analysis of the benefits of this requirement was exhaustive. The agency detailed a substantial net energy gain from recycling a quarter of the waste stream relative to incinerating it. It detailed significant expected reductions in the toxicity of air emissions and the associated public health risk from incinerators; certain wastes that produce toxic by-products when incinerated would be diverted (primarily plastics, but also household batteries, which contain heavy metals and were specifically required to be removed), and the total size of incinerators would be reduced. It cited significantly improved combustion efficiency and a large reduction in the amount of incinerator ash—a hazardous waste by most traditional measures although it had yet to be so regulated under RCRA.[37]

The large companies that dominated the field of incinerator construction quickly expressed their opposition in formal comments on the proposed regulations and through pressure applied on "Business's Backdoor Boy," as then Vice President Dan Quayle, chair of the President's Council on Competitiveness, had been labeled.[38] Foster Wheeler Power Systems, Wheelabrator Technologies, and Ogden Martin Systems argued that the EPA did not have authority to require recycling (by requiring its precursor, materials separation) under the Clean Air Act; that, as David Littell summarized the position, "a materials separation requirement pursues a non-air quality purpose that is not valid under the Clean Air Act."[39] It was just the latest variation on longstanding efforts by multiple industries to keep recycling in a postconsumer waste box, decoupled from broader environmental objectives in pollution prevention, energy conservation, and resource protection.

In February 1991, the EPA published the final rules, which, in a complete reversal of its earlier argument, eliminated the source separation requirements. The crux of the new argument ran along the following lines: the actual reuse of the separated materials couldn't be assured because of market uncertainties, so the various benefits earlier proposed were indeterminate; some recycling would probably occur anyway; and emissions standards for incinerators was an inappropriate vehicle for solid waste planning. As Karen Kendrick-Hands wrote, "Notwithstanding all these rationalizations, the real reason the Environmental Protection Agency backed down on the twenty-five percent materials separation requirement, so warmly espoused in the proposed regulations, and the proposed Senate version of the Clean Air Act [reauthorization], was pressure from the Bush White House, including the President's Council on Competitiveness."[40]

In one stroke, the decision overtly and unequivocally summed up the central themes of the federal solid waste policy as they had played out for more than a decade: obstruction of recycling and support of incineration. To find reasons that can better explain the significant growth in both wastepaper recovery and utilization, and models for future efforts, we must look elsewhere.

Lessons from Newsprint

The recovery and reuse of newsprint is of particular interest in the assessment of domestic paper recycling progress, partly because its recent history is relatively well documented and because some of the lessons of newsprint recycling may be generalized to other wastepaper and product classes. It has also long been the most visible of all wastepaper recycling efforts; newsprint had been a mainstay of the small recycling programs that existed before the recycling movement of the 1980s, and it was the first material to be targeted in the new curbside collection programs that began to proliferate. It is, like corrugated containerboard, an easily recovered postconsumer wastepaper grade because it is primarily made into newspapers, which can be clearly identified and separated at the source. Throughout the late 1980s and early 1990s newsprint recycling became the most prominent symbol of both the successes and failures of the modern recycling movement, as well as a focal point for many of the state efforts around recycling.

By the early 1990s, success was conditionally proclaimed in the many lists that began to be regularly published detailing the new newsprint deinking projects that were being proposed and coming on line. However, it followed, not coincidentally, on the heels of a rather dramatic and highly visible failure: the total collapse of the ONP market in 1989. By year-end 1989, ONP prices in many regions were below $0; recycling program managers quite literally could not give the stuff away, and the conditions of severe oversupply were expected, at the time, to persist until the mid-1990s.[41]

The popular explanation for this troubling state of affairs was that supply-side recycling approaches, primarily the mandatory wastepaper collection programs put in place by a variety of states, counties, and cities, had failed to be accompanied by either solid market research or demand-side policies such as government procurement standards. Indeed, much of the legislation put in place at the state level and below as the ONP oversupply problem worsened was focused on the latter. In California, for example, AB 939, a 1989 state law that required cities and counties to prepare integrated waste management plans and to meet designated targets for the diversion of municipal waste to recycling, was soon followed by AB 1305, which required newspapers published in the state to have a minimum level of recovered fiber content. Many other states followed a similar two-step approach, with increased materials recovery as the first step and "market development" (broadly defined) as the second. Some favored minimum content standards like California's; others took a softer approach by negotiating voluntary content goals with newspaper publishers and/or offering tax incentives to publishers if they reached minimum content goals; while still others approached the problem by proposing minor taxes of various sorts on virgin newsprint.

A vast assortment of government procurement programs were put in place at state and local levels, although they had somewhat limited relevance to newsprint. Some states revisited the supply-side issues, developing mandates and incentives focused on a more sophisticated collection infrastructure and cooperative materials collection and marketing plans. Wastepaper export development was promoted and pursued at almost every level. A few cities and eventually some states made a tentative break with the black box approach to the industrial variable in the equation and undertook limited attempts to stimulate investment in recovered materials-based manufacturing capacity by offering various tax credits for the purchase of recycling equipment and by proposing recovered materials-use credits (many of which were equally applicable to investment in waste-to-energy incinerators.)[42]

It is difficult to estimate which of these market development-oriented policies had the most effect in terms of stimulating the more than 1 million tons per year of new ONP-using capacity that came on line in the United States between 1990 and 1994. Indeed, the sea-change in recovered newsprint utilization (particularly in newsprint production) beginning around 1990 could be linked to a variety of factors, many of which had surprisingly little to do with market development efforts. As one analyst reviewing the recent volatile history of ONP and newsprint questioned in 1994, "Is it supply that drives demand, or demand that drives supply?"[43]

In addition to a dramatic increase in ONP recovery resulting from mandatory collection programs, however, there was another story behind the collapse of the ONP market. Although the events transpired with remarkably little notice by the recycling community, in 1989–1990—at the height of national interest in recycling and as the problems of ONP oversupply became front page news—three new virgin newsprint mills opened in the United States, adding a total of more than 700,000 tons per year of virgin woodpulp capacity.[44] The incongruity of opening three virgin newsprint mills at a time when the country was literally awash in recovered ONP was symptomatic of an underlying failure buried in the changing trade balance for newsprint during the 1970s and 1980s and in the lost opportunities of those decades. There was an inner logic to the growth and trajectory of newsprint production during those decades that in fact proceeded with remarkable indifference to both national recycling interests and to a set of market forces commonly considered to affect positively the development of industrial recycling capacity.

Newsprint supply differs markedly from the supply of other grades of paper in that the United States has long been heavily dependent on Canadian producers for a large proportion of its newsprint consumption. Throughout the 1960s, domestic newsprint production represented less than a third of total newsprint consumption. From the 1960s forward, however, domestic newsprint production consistently rose at more than twice the rate of growth of newsprint consumption as domestic producers sought to claim a larger share of the huge U.S. market. Between 1970 and 1990 newsprint production grew at an annual rate of 3.5 percent and increased by 3.4 million tons (to 6.6 million tons; see table 4.4).[45] Domestic producers' share of the domestic market rose from 34 percent in 1970 to almost 50 percent in 1990. For more than two decades prior to 1989, newsprint capacity had been steadily expanding and had grown more rapidly than total production in the industry.

By the late 1980s, however, only 30 percent of all recovered newsprint was being used in newsprint production. The remainder was being used in other paper products (40 percent) and miscellaneous nonpaper applications (10 percent), and more than 20 percent was being exported. Although newsprint can be produced with up to 100 percent wastepaper content, the wastepaper utilization rate in domestic newsprint production was about only 24 percent in 1988. Given the large supplies of virgin newsprint coming in from Canada, essentially all of the substantial expansion in domestic newsprint production—a 2.6 million ton increase in annual production between 1970 and 1988—could have been based on recovered materials. That less than a third of it was, and that domestic manufacturers were still expanding virgin newsprint production capacity by the late 1980s, represent further complications in the argument that basic market forces provide the strongest and most efficient stimuli for recycling.

Throughout the 1970s and most of the 1980s, in the absence of any serious state or federal intervention to increase paper recycling, the basic economic conditions that might generally have been expected to support significant investments in increased recycled newsprint production included the following: sufficient or excess supplies of wastepaper, low wastepaper costs relative to woodpulp costs, favorable process costs associated with using wastepaper, the availability of established newsprint deinking technology, favorable capital costs associated with increasing deinking capacity versus increasing virgin pulp capacity, and declining availability of wood. During this period, all of these basic market forces, with the exception of a significant scarcity in wood supplies at the national level, were present to a heightened degree.

Table 4.4
Newsprint production, wastepaper use and recovery, 1970–1994 (million tons)

	Newsprint production	Newsprint consumption	ONP* recovery	Net ONP exports	ONP use all mills	ONP use newsprint mills	Other wastepaper use newsprint mills
1970	3.35	9.80	2.49	0.05	2.24	0.37	0.00
1975	3.48	9.05	2.48	0.14	2.04	0.46	0.00
1980	4.70	11.38	3.28	0.42	2.56	0.85	0.05
1985	5.43	12.84	3.87	0.64	2.88	1.31	0.00
1986	5.63	12.99	4.29	0.77	3.12	1.36	0.00
1987	5.84	13.71	4.49	0.90	3.14	1.39	0.00
1988	5.98	13.67	4.75	1.04	3.22	1.43	0.00
1989	6.09	13.10	5.40	1.28	3.70	1.50	0.01
1990	6.61	13.30	6.00	1.26	4.20	1.80	0.10
1991	6.84	12.50	6.70	1.30	4.60	2.00	0.20
1992	7.08	12.60	7.10	1.31	5.10	2.30	0.20
2000	8.50	14.70	10.10	2.02	6.90	3.40	0.90
			High export scenario:	3.01			

*As used here, ONP includes postmill/preconsumer recovered newsprint.
Sources: Franklin Associates, Ltd., *Paper Recycling: The View to 1995* (1990) and *The Outlook for Paper Recovery to the Year 2000* (1993); *Lockwood-Post's Directory* (annual) (1970s production).

The recovery rate for newsprint had been increasing steadily throughout the period, from about 25 percent of consumption in 1970 to almost 36 percent by 1988. Although much of this growth was going to exports, the general momentum and infrastructure for increasing recovered newsprint supplies had been steadily building. In any event, the availability of recovered newsprint as a fiber resource, even on a regional basis, had never been a serious limiting factor in its use in newsprint production due to the huge gap between domestic production and consumption. Although the low cost of wastepaper had to be measured against the security of the supply infrastructure, for newsprint the facts that consumption was more than double production and that collection was comparatively easy rendered the security of the supply more assured than for any other grade. Throughout the period recovered newsprint was both plentiful and cheap.

With respect to the comparative costs of recycled newsprint production, as early as 1973 Franklin had reported, "our estimates show that under present conditions a typical domestic newsprint mill probably incurs manufacturing costs of $20 to $30 more per ton than a newsprint deinking mill running at full capacity."[46] The energy conservation advantages of recycled newsprint production had been well understood since the early 1970s and had never been seriously contested nor subject to the same analytical complexities that attended energy use comparisons between recycling and chemical pulp production. Unlike chemical pulp mills, which derive a large proportion of gross energy use from the combustion of pulping liquor, high-yield mechanical pulping processes (including those used to make newsprint) are, like recycling processes, dependent on purchased energy.

Although there had been a surge in public enthusiasm for recycling in the 1970s, in the aftermath of the first Earth Day and again during the energy crises of the late 1970s, and although apparent economic conditions seemed optimal for a significant increase in wastepaper utilization in newsprint, recycling efforts nevertheless largely stalled. Neither the industry nor government provided effective follow-up to the initial burst of enthusiasm, and the economic conditions favorable for a substantial expansion of recycled newsprint production appeared insufficient to substantially overcome other forces influencing industry expansion.

The capital and production costs of using recovered materials in newsprint manufacture had consistently appeared favorable, but other factors influencing industry decisions included: long-term fixed contracts between mills and newspaper publishers, or the ownership of newsprint mills by publishers; a general bias against recycled paper; and the industry's longstanding identification of itself as a wood-based industry plus the traditions and internal culture that accompany such a perception. Combined with the realities of extensive structural integration within the forest products industry and the elaborate infrastructure, including hundreds of regional and national trade and lobbying organizations, built up to guarantee its access to wood supplies, these factors worked against ONP utilization.

Nevertheless, prior to the height of national interest in recycling in 1989–90 and the ONP glut of those years, recycled newsprint capacity did increase. If, at a national level, it was not at a rate that might have been expected due to the highly favorable market conditions for expansion, the divergence between the major producing regions was difficult to miss. Throughout the 1980s, although the conflicts would remain extreme and would not peak until the end of the decade, the tide was turning for the forest products industries in the West, and the shift to the South had already begun. In fact, most of the new recycled newsprint capacity added prior to 1989 was located in the West. By 1988, although only 30 percent of domestic newsprint production was based in the West, the region accounted for nearly half of all ONP deinking capacity. The South, which accounted for almost 60 percent of domestic newsprint production, had less than 20 percent of domestic ONP deinking capacity.[47] Wastepaper availability and recycling cost factors being more or less equal, the pending threat of timber scarcity in the West seemed finally to drive the wastepaper utilization level. As one Western newsprint manufacturer pointed out as late as 1990, the increases in recycling capacity in northwestern mills were "a hedge against what will be happening in the future with the timber supply."[48]

If the shock that reverberated around the ONP glut in the late 1980s—colliding with the high public expectations around recycling—seemed to catch the domestic newsprint industry by surprise, the mandatory minimum content standards quickly imposed by several states caught Canadian newsprint producers completely off guard.

In the United States, Garden State Paper had been producing recycled newsprint since the 1960s, and seven or eight other domestic manufacturers had also begun producing recycled newsprint. Most prominent was the Jefferson Smurfit Corporation, which, on the West Coast alone, was producing more than 750,000 tons per year of recycled newsprint in two large mills in Oregon and one in southern California. Whereas the wastepaper utilization rate in domestic newsprint was around 24 percent by the late 1980s, in Canada it was about only 2 percent. Although most newspaper publishers and newsprint manufactures argued against minimum content standards, especially in the Northeast (which was particularly dependent on Canadian producers) and in the South (which had substantial new virgin newsprint capacity), protests by both manufacturers and publishers in the West were less forceful. Indeed, several western publishers were quick to proclaim the advantages of recycling, including the *Los Angeles Times* (which had a stake in the Smurfit Newsprint Corporation), and the *Seattle Times* (located in a region with growing timber problems and the most sophisticated recycling infrastructure in the country). In fact, the president of the Seattle newspaper went so far as to deliver a lecture—actually a scolding—to the Canadian forest products industry:

> My newspaper—a large consumer of Canadian newsprint—and many other newspapers in the States are perplexed by the failure of the Canadian industry to respond to the irreversible stampede toward using recycled newsprint . . . [S]ome environmentalists in the States are comparing the people who run your forest industry to those who are raping the forests of Brazil. And, unfortunately, newsprint mills in Western Canada are doing little to disabuse this kind of criticism. Instead we hear excuses why recycling won't work.[49]

Of course, other environmentalists were comparing the people raping the forests of Brazil to those who run the forest products industry in Washington and elsewhere in the western United States.

The tone of this lecture would later be echoed in the chastisement of U.S. producers for recalcitrance on TCF bleaching, delivered by the more advanced and more advantageously situated Swedish industry. California's standard—which affected more than 15 percent of domestic newsprint consumption—was soon followed by similar proposals in Oregon and Washington and provided what amounted to a competitive edge for Western newsprint manufacturers relative to their counterparts in British Columbia. This advantage derived from the already established base of recycled newsprint production and from the existing and expanding collection infrastructure driven by a solid waste crisis that had yet to develop to the same extent in Canada.

The factors that have driven (or failed to drive) expansions in recycled newsprint production are complex and have often differed substantially by region. However, the extremely bad press associated with the collapse of the ONP market began, for the first time, to bring the industry itself into sharper focus. As one newspaper publisher put it, "Legislators, environmentalists, and others began screaming that the newsprint industry—and their customers, the newspapers—were not acting responsibly and were not willing to make and use recycled newsprint."[50] By 1990 the paper industry was struggling to undo the damage done and to recast itself as a partner in the national recycling challenge, although its efforts in the early 1990s would often be characterized by conflicting pronouncements and policies. Despite a wastepaper recovery rate that in the late 1980s was growing by leaps and bounds due to crisis-driven mandatory state collection programs, the industry expressed its commitment to the challenge by announcing a superfluous national wastepaper recovery goal. It then, in an astonishing revision of history, attempted to recast itself as the central agent in the increasing wastepaper recovery rate. The president of the American Paper Institute testified in 1991, "The infrastructure now in place . . . is the result of decades of work by the industry with communities in developing collection systems."[51]

At the same time, the industry continued to demonstrate a posture toward solid waste management that ran directly contrary to policies that produce more assured municipal wastepaper supplies. The API, for example, continued to call for "integrated" waste management that "*balances* the principal options of market-driven source reduction, recycling, waste-to-energy, composting and landfilling" (emphasis added). (One would be especially hard put to find any reason for thinking that source reduction would be driven by market forces in the foreseeable future.) In particular, the trade association called for "greatly expanded use of waste-to-energy technologies" (i.e., incinerators).[52] The industry also tended to express its quantitative connection to the use of wastepaper in only vague terms—calling for recovery of 40 percent of domestic paper consumption for "domestic recycling and export," and describing progress in terms of lists of projects and recovery rates rather than wastepaper utilization rates.

At the periphery of the core industry, however, this position—which was attempting to keep the heat off the industry and to avoid a repetition of the ONP debacle by keeping all other options, from incineration to export, open—began to moderate. It provided the first indication that real progress and a long-term commitment to a higher level of recovered fiber use had finally begun to take hold. The Paper Recycling Coalition, an association of ten small paper manufacturers producing 100 percent recycled paper products, sounded a new note in 1990 testimony: "While we recognize that there will be an increase in the export of recovered materials, it is critical that the recovered materials and fiber needs of the domestic industry continue to be met."[53] This not only broke from contemporaneous API pronouncements, but also contrasted with the "market mania" of the then lively RCRA reauthorization debates in which interest in recovered materials export development was stronger than ever.

Another sign that the industry had begun to pursue more than a public relations angle in its activities around wastepaper came—most strongly from the Paper Recycling Coalition, but echoed by the API—in one of its priorities for RCRA reauthorization: "[R]ecovered materials should be excluded from the definition of solid waste and remain completely outside the authority of solid waste regulation." While much of the rest of the recycling community was still catching up with the issues of market development, the industry's attention was becoming focused on the mechanisms of collection.

Wastepaper collection had been renewed and then stalled in the 1970s at the level of small-scale, largely voluntary community efforts, which Louis Blumberg and Robert Gottlieb have described as less a business proposition than "an extension of social and cultural movements . . . motivated largely by concerns about a conservation 'ethic' and the proliferation of roadside litter."[54] By the 1980s, however, as the garbage problem worsened and larger-scale programs were put in place, the small programs and entrepreneurial recovered materials brokers began to be either absorbed or displaced by large waste haulers such as Waste Management and Browning-Ferris. Increasingly, large paper companies began to acquire divisions and subsidiaries for secondary materials collection and procurement development, or they cautiously entered into joint ventures with the large waste hauling companies. Efforts to distinguish wastepaper recovered for recycling from wastepaper collected for landfilling or incineration thus represented an attempt both to avoid unwieldy RCRA-style regulation of the transport and interstate shipment of wastepaper and to establish control of what was becoming a more interesting commodity. As the Paper Recycling Coalition asserted,

> Materials which have not been discarded, and which have been diverted or separated from the solid waste stream for purposes of recycling, are not solid waste and are not subject to flow control as solid waste. *Ownership of such materials cannot be 'taken' by government,* either directly, or through [other] restrictions . . . (emphasis added)[55]

Another new theme in the discourse could be discerned in relation to some of the smaller-scale mill developments and was suggested by the Paper Recycling Coalition, which described itself as being largely composed of "small, family-owned" businesses. It seemed in part to be a sign of the famed American entrepreneurial spirit, so often belabored, but so little in evidence in the paper industry sphere since the last century. Emerging articles on new small-scale recycled tissue mills reflected this quality: "Persistence, Common Sense Are the Qualities Behind Ponderosa's Success" claimed one article. "It was mid-January when the project got underway," observed John Shultz, in an article on his new tissue minimill near Memphis, Tennessee, "the construction crews, with their no-nonsense attitude, were out there digging the foundations in the cold and the rain, pumping out water, and going back to work." This new breed of paper entrepreneur evoked a sharp contrast with the multimillionaire executives of the multibillion-dollar paper companies.[56]

At the same time and in conjunction with the development of these smaller projects, recycling had come to be defined as an economic development opportunity—particularly for the troubled urban centers of the country. Indeed, the pursuit of recycling-related development had, in the early 1990s, become one of the centerpieces of the environmental community's delayed response to the damaging "jobs versus environment" conundrum advanced and nurtured by corporate America as well as of rising attempts to provide substance to the rhetoric of "sustainable development" and "sustainable economies."[57]

The positive potential of this view, which had become a new movement in its own right, continues to be significant. However, if the framework continues to be narrowed, duplicating and building on a central failure of the 1980s recycling movement, the negative potential is also significant as reflected in the following chapter. On the positive side, small-scale mills using not only urban waste, but also potentially other suitable regional materials, may indeed help to recapture some of the employment lost in the concentration of the paper and other core manufacturing industries, to revitalize the prospects for otherwise abandoned urban and rural populations, to meet multiple environmental goals, to reintroduce a visible connection between production and consumption, to establish an improved level of community equity in production, and to mark the first step in a transition to a paper industry that is more decentralized and flexible. Brenda Platt, of the Washington, D.C. Institute for Local Self-Reliance has argued, "Processing scrap may not sound as jazzy as building a convention center or soaring office buildings . . . but it promises to have a more profound impact on the local economy."[58] The easy path, and a sure path to failure, however, is to lose sight of the connections to the extremely powerful industry on whose periphery these prospects sit, to the inflated levels of consumption that have been driven by an industry that is the archetype of mass production, and to the condition of the natural resource base that supports paper production.

This chapter has attempted to sketch out the broad dimensions of some of the forces that have operated both within and upon the paper industry with respect to past and prospective recycling progress. It has also attempted to suggest a more meaningful way—a way that begins to reflects multiple goals—to quantify and monitor real progress. Even if we measure and monitor timber and woodpulp production and trade as part of our paper recycling analyses and policies, and even if we more fully consider the scope of opportunity for recycling and alternative fiber use by the paper industry as part of our forest policies or as part of pollution and energy policies, there are ultimately no guarantees that recycling can be made to have a significant impact on these issues. What we *are* guaranteed by history is that if we ignore these considerations it almost certainly will not.

Left to market signals that provide little challenge to the inertia of the core industry and long-held economic relationships and that fail to reflect the real costs of timber production, solid waste disposal, or environmental pollution, the industry will either stand still, or move forward so slowly that it continues to be outpaced by the rate of growth in the problems, or change only slightly at the margins. In failing to face the larger issues, we have instead often used the blunt tools of federal and state policy on the micromanagement of isolated parts of the problem. There is a difference between intervention focused at the heart of the complexities of market allocation of wastepaper resources and intervention focused on making a large chunk of new resources available or a large chunk of traditional resources unavailable and on defining basic national objectives.

Arguably, despite the failures of market assessment and long-range planning and despite all the other policy failures that led to the ONP glut and angered a public that had so optimistically embraced recycling, the collapse of the ONP market had one of the most positive impacts on recycling progress seen in recent decades. The disruptions it caused may have been a small price to pay to begin to awaken policymakers and the public to the centrality of the paper industry's role in the national paper recycling equation. Prior to 1989–90, however, the clearest force that stimulated wastepaper utilization in newsprint production, echoing the effects of the resource scarcities of the war years decades earlier, was the inexorably shifting forest policy in the Pacific Northwest.

Although headway in recycling has been made since the early 1990s, solid waste and energy crises, crashing wastepaper markets, public anger, regional resource conflicts that escalate to levels of violent polarization, and wars have historically represented the most effective stimuli to recycling. As the following chapter suggests, this conflict-driven approach has continued to dog modern progress in paper recycling in sometimes unexpected ways.

5

Case Study of an Urban Recycling Project

In August 1991 Norm Masters, president of Quality Urbanization and Development (QUAD), a community group based in West Sacramento, California, met with Dan Ramos, a local industrial property developer. Over lunch the two discussed issues surrounding the rerouting of local rail lines; one of many local planning issues on which their opinions distinctly differed. As the meeting ended, however, Ramos remarked offhandedly "Boy, do we have a project that you guys are going to love!" The project he was referring to was a proposal by MacMillan Bloedel, a major Canadian forest products company, and Haindl Papier of Germany, to build an integrated recycled paper mill on an undeveloped property known as the Southport Industrial Park (a property in which Ramos's father was a key investor). Although discussions between the Southport investors, the paper companies, city officials, and environmental regulators were already well underway, it was the first time that Masters—a long-time resident and avid community watchdog—had heard about it.

If built as proposed, the mill would consume ONP, OMG, and OWP, and produce 100 percent recycled newsprint along with some recycled printing and writing grade paper. Yet, unlike the urban minimills that had begun to appear in a few places, this facility—representing the logical culmination of a set of forces of which the community had little inkling—would be the largest recycled paper mill that had ever been built and one of the largest pulp and paper mills in the world. In this sense, it represented a merger of one of the most promising opportunities for the paper industry with one of its worst tendencies: bigger is better.

The environmental review and permitting processes surrounding this proposal would preoccupy Masters and other members of QUAD for the next eighteen months and would still be unresolved at the end of that period. The project would also incur the animosity of most of the regional environmental community, including air and water quality regulators as well as the state's Energy Commission, and it would come to be opposed by the regional agricultural community and by constituencies as seemingly distant as the powerful Metropolitan Water District in Los Angeles. However, the national coverage the conflict ultimately received focused on the antibusiness tendencies it was said to represent. In one particularly vitriolic editorial, *Forbes* magazine called project opponents a "lynch mob" and accused state regulators of believing in the "tooth fairy."[1] The conflict cast into sharp relief a set of issues that had begun emerging around urban recycling development in other places, although never—before or since—in such dramatic proportions.

The simplistic perspective of national business observers notwithstanding, the real issues revolved around the appropriateness of the scale and form of new industrial paper recycling capacity within the context of both local and statewide environmental and economic objectives. They highlighted the different geographic scales at which the issues needed to be understood, as well as the difference between hypothetical net gain at the larger scales and concrete local reality. The problems surrounding the proposal would call into question certain core assumptions about recycling and would quickly obsolesce the solid waste view. The depth of the conflict could ultimately be seen as symptomatic of the vulnerability of recycling policies that are blind to the dynamics of the virgin materials industries. It would also become clear that there was no single path by which to expand recycling capacity, yet that the existing policy frameworks embodied a one-dimensional view that failed to distinguish between the alternatives.

The process surrounding the environmental review of the facility, which is the subject of this chapter, provides a concrete illustration of the intersection of recycling policies framed by solid waste management concerns, with a larger set of environmental and industrial issues. In this case, the intersection can most accurately be described as a head-on collision.

A Perfect Setting?

West Sacramento is a young city, incorporated in the 1980s, with a 1991 population of approximately 28,000. It is located near the center of California's Central Valley, immediately adjacent to the capital city of Sacramento. The Sacramento metropolitan statistical area includes El Dorado and Placer Counties in the Sierra foothills, as well as Yolo County (in which West Sacramento is located) and Sacramento County (in which the City of Sacramento is located). It has a population of approximately one million. Throughout the late 1980s it was the fastest growing metropolitan area in the nation, although by 1991 the region, like the rest of the state, was mired in an intractable recession, with an official unemployment rate above 9 percent.[2] A middle-class community of trade, technical, and service workers, retail workers, state government employees, educators, and small business owners, West Sacramento is characterized by modest two- and three-bedroom houses on generous lots built between the 1940s and the 1960s, with newer residential tracts and commercial developments rising in its undeveloped, formerly agricultural areas. Along its deep water shipping canal and the Sacramento River, it is dotted with older manufacturing facilities and related infrastructure.

Like many incorporation efforts, West Sacramento's had been driven by a desire to gain greater local control during a period of rapid regional growth: the community had increasingly begun to feel as if it were being viewed as a dumping ground for land uses resisted by powerful neighboring cities. Upon incorporation, however, the city had inherited a number of land-use agreements, including the Area Plan and Environmental Impact Report for the Southport Industrial Park. After a series of bitter fights, these documents, certified by Yolo County in the early 1980s, were retained as part of the city's general plan with little modification. The city has a core of long-time residents, many of whom had participated in the incorporation effort and the more recent battles over the general plan and who came to form the nucleus of the QUAD membership and ultimately of the opposition to the paper mill. It is worth noting that QUAD did not view itself as an environmental group, and most of its members would have been uncomfortable with the label. It was also not antigrowth or antibusiness (its members included some local business owners) but rather was concerned with precisely what its name implied: quality urbanization and development.

The city is also located at the heart of California's water system in an area known as the Sacramento-San Joaquin Delta. Sacramento sits at the confluence of the Sacramento and American Rivers, which drain the northern Central Valley and the northwestern Sierra watersheds. South of the city, the Sacramento River is joined by the San Joaquin River, which drains the central and southwestern Sierras. The surrounding area is largely reclaimed marshlands protected by an elaborate system of levees and dikes, and it contains some of the most productive agricultural land in the world. The Delta water system flows to the San Francisco Bay some eighty miles to the west and has been at the center of water controversies in California for decades because the ecosystem has been profoundly degraded and imperiled. Major impacts on the ecosystem have come from freshwater withdrawals and diversions used to supply imported water for both agriculture in the arid central part of the state and southern California cities. The Delta is also stressed by pollution, including pesticide and fertilizer contaminated agricultural runoff, plus municipal and industrial wastewaters.

Despite the fact that the sensitive ecology of the Delta would inevitably become an issue in planning and permitting a large industrial facility in its watershed region, the Sacramento area offered a number of distinct advantages from the paper companies' point of view. It had an ample supply of fresh water and fewer air quality constraints relative to the San Francisco Bay area and Los Angeles, two other candidates for paper recycling development in California. (Actually, the region has a steadily worsening air quality problem; however, the issue had not reached the level of volatility nor had the local environmental community reached the level of organization that characterized these other areas.) The Sacramento region also had a sizeable local supply of wastepaper, as well as easy transportation access to the Bay Area, other Central Valley population centers such as Fresno and Modesto, and potentially the enormous but more distant population of southern California.

The flip side of huge wastepaper supplies is, of course, a huge paper market. For MacMillan Bloedel in particular, it was not only an opportunity to access new markets but also a question of maintaining its existing California markets. There, the company already had sales of more than 1,000 tons per day (or the total daily output of a large newsprint mill) of paper produced at its mills in British Columbia.[3] California, however, had recently passed solid waste legislation that mandated municipal solid waste recovery for recycling (AB 939, the California Integrated Waste Management Act of 1989), as well as legislation setting minimum wastepaper content standards for California newspapers (AB 1305). The latter, in particular, posed a problem for MacMillan Bloedel. To meet these new standards, the company would have to dramatically increase its use of Canadian wastepaper to supplement its virgin operations in British Columbia or find another alternative. Although the company had become involved in a small recycled market pulp facility in Coquitlam, British Columbia (which had itself been a source of controversy) the depth of the MacMillan/Haindl commitment to the California-based strategy was indicated by their proposed investment of $1.5 billion in the project.[4]

Details of the project did not become publicly available until the first draft Environmental Impact Report (EIR), required under the California Environmental Quality Act (CEQA), was released in mid-January 1992.[5] At that time, it became apparent that the facility would engage in an unprecedented scale of production. The patterns of non-disclosure and manipulation of regulatory process that would come to characterize the approach of the developers and the city council were also first revealed. Although continually cast in the high rhetoric of an environmentally and economically beneficial development and shielded to some extent by the positive public perception of "recycling," this was not to be a project planned in consultation with the community, nor was it by any measure sensitive to the local environment.

The project was designed to be built in three phases over a period of fifteen years. Each phase would consist of one or two wastepaper pulping lines employing conventional flotation deinking processes, combined with a paper machine. Phases I and III would each include a 46 MW gas-fired cogeneration plant (an industrial boiler combined with a steam turbine) that would produce both electricity and steam for industrial processes and have ninety-foot-high discharge stacks for air emissions. An additional component of Phase I (later isolated as "Phase Ia") would consist of what the project proponents called a "solid-waste-fired steam generation facility," but project critics identified as a "sludge incinerator." It would be augmented by a steam turbine in Phase II to cogenerate an additional 10 MW of electrical energy as well as steam for Phase II operations.

The facility would require 3 million gallons per day (mgd) of fresh water for each phase, for a total of 9 mgd (or nearly 10,000 acre-feet per year), and would generate 7.8 mgd of industrial wastewater. The wastewater would be treated at a new municipal wastewater treatment plant to be financed and built by the city and then released to the Sacramento River. At completion, the project would also use more than a million pounds of process chemicals per day, or about a quarter ton of chemicals per ton of product. Chemicals would include known hazardous chemicals such as sulfuric acid (more than 16,000 lb./day) and anhydrous ammonia (2,550 lb./day), as well as unidentified chemicals in products such as "Slimetrol" slimicide and felt cleaner. (They would not, however, as prominently advertised in company literature on the project, include chlorine.)

If establishing a market for California's rapidly increasing supplies of recovered wastepaper were the only public policy consideration, then there was no question that this project was exactly what the state needed. At completion, the mill would consume nearly a million tons of wastepaper per year. Specifically, Phases I and II would each consume about 990 tpd of wastepaper consisting of ONP (63 percent), OMG (27 percent), and office waste (10 percent); and each would produce 790 tpd of 100 percent recycled newsprint. Phase III would consume about 560 tpd of wastepaper and 115 tpd of purchased bleached kraft market pulp, and it would produce 555 tpd of recycled office-grade paper.[6] California's 1991 consumption of newspaper and magazine products was estimated at 2.3 million tons, and projected to increase to between 3.1 and 3.8 million tons by 2010. Under these scenarios, the project would eventually be consuming between about a quarter and a third of total generated ONP/OMG in the state. In 1991, California was exporting about 78 percent of its recovered ONP/OMG to other North American locations (40 percent) and to off-shore locations (60 percent).[7]

The calculations did not include an analysis of the impact on total office wastepaper generated or recovered. In general, there appeared to be a de-emphasis of the office waste and OMG consumption and of the plans for recycled printing and writing grade paper production in Phase III. The more intensive deinking processes expected to be required to produce higher-grade papers raised potentially more serious implications for solid waste, air emissions, and effluent quality—particularly troublesome in the context of the volatile Delta water quality issues. However, wastewater, sludge, and ash samples from "comparable" operations at other facilities—the basis of permit applications and the EIR analysis for this project—had not included samples from deinking processes producing white-grade pulps.[8]

Although a substantial market would be created for California wastepaper, the facility would also produce about 55 tpd of dry solid waste and at least 1,000 tpd of wet sludge for a combined total of roughly 370,000 tpy. How to manage this waste, including whether or not to incinerate some of the sludge and whether the sludge or incinerator ash would be designated as hazardous waste, would become one of many controversies dogging the environmental review process. It quickly became apparent that solid waste from the project, with or without incinerator ash (which might require a separate ash monofill to be built), would have a severe impact on local landfill capacity.

Quantitative estimates of the solid waste impact on area landfills were constantly shifting with each new scenario for managing the waste. It was variously proposed that the waste would be disposed at one or both of two Class III (municipal waste) landfills in the area; that it would piggyback on possible future long-haul waste transport from San Francisco to "a remote disposal site;" or that it would in part be used as landfill cover material, as an agricultural soil amendment, or as a supplement to municipal composting projects (proposals that would require pilot projects and rigorous testing before they could be permitted because the toxicity of the waste had not been ascertained). The ash from the incinerator might be disposed in a Class III landfill or in a separate Class II (hazardous) monofill to be sited either independently, adjacent to a Class III landfill, or adjacent to a Class I (extremely hazardous waste) facility, and so on.

In any case, the most probable scenarios for the disposal of solid waste from the project appeared likely to shorten the life of the Yolo County landfill by at least ten years, even with the incinerator. The irony of producing a serious landfill capacity problem by building a recycling plant was not lost on area residents. Perhaps even more clearly than the air and water quality issues, which were highly technical and plagued by inadequate test data and the complexities of risk assessment and regulatory process, this irony highlighted the fact that the community was being asked to accommodate a vastly disproportionate share of the state burden to increase recycling levels. Although the wastepaper would be acquired from an area with a radius of up to 300 miles (thus spreading the solid waste diversion benefits across all of northern and central California), the negative environmental impacts of production would be largely concentrated within a radius of perhaps 5–10 miles in an urban and suburban setting.

At completion the production level would be more than 2,100 tpd: about two to three times the size of most of the large integrated virgin newsprint mills operating in North America, most of which are operating in lightly populated rural settings. Indeed, at a production level of roughly 750,000 tpy, the mill would be competitive in size with the very largest kraft woodpulp mills in the world. It is safe to say the facility would fall in the range of five to ten times larger than most existing urban paper mills. The Smurfit newsprint mill in Pomona (a suburban industrial area in the Los Angeles basin), for example, the only other newsprint mill operating in California, was producing about 420 tpd of 100 percent recycled newsprint from wastepaper deinked on site. Similarly, the Newstech recycled market pulp mill (also located in a suburban area) with which MacMillan Bloedel was associated in British Columbia was also operating at a level of only about half that proposed for Phase I alone of the three-phase West Sacramento project. Because there were no precedents even remotely comparable to an urban mill of this scale, the detailed environmental analysis required by CEQA became particularly crucial.[9]

Economic Blackmail and Regulatory Shenanigans

The initial draft EIR employed a confusing scheme that broadly outlined the three-phase project, but provided details on only the first phase, making it unclear exactly what project was being proposed. Its forty-five day public review and comment period was barely underway before the city decided to produce a revised draft of the document. The revised version, released in May of 1992, was substantially improved by addressing all three phases of the project, but nevertheless raised almost as many concerns as it answered. Almost none of the deluge of problems and detailed questions raised in written comments and public testimony by individuals, community groups, environmental organizations, downstream farmers, water contractors, other cities, and state agencies were satisfactorily addressed or resolved in the final EIR, which was released in August.

The approaches taken in assessing and proposing mitigations for the air and water quality impacts were particularly troublesome. One factor complicating the discussions was that the city repeatedly claimed that the project would be in compliance with all applicable environmental regulations and that this was for practical purposes equivalent to a finding of "no significant environmental impact." This created predictable problems. The first was that "significant" is clearly a relative term, open to both subjective interpretations and various legal interpretations established under different environmental statutes. A significant impact from the point of view of city residents or downstream farmers was not necessarily the same as a significant impact under the terms of CEQA, which in turn did not necessarily correspond to interpretations of significance under air quality or other pollution control laws. A second problem was that final determinations on environmental permits had not been made by the responsible environmental agencies, and much of the required sample testing and analysis had not been completed. The EIR presented sketchy and inadequate quantitative data on selected pollutants from selected operations, but provided no comprehensive profiles. When questions were raised concerning specific pollutants, the city consistently responded that environmental regulations would be observed, but in most cases it failed to supply any further detail or to substantiate its claims. The final and most serious problem, as many respondents were aware, was that there is no guarantee that compliance with environmental regulations will produce an environmentally sound project. This was particularly at issue in terms of air quality requirements.

Although the federal government and authorized state air pollution control agencies had long regulated the six criteria air pollutants, they had long failed to effectively regulate toxic air pollutants. The 1990 Clean Air Act amendments attempted to address this failure by identifying 189 hazardous air pollutants to be regulated under a two-phase strategy. However, the federal regulations would not be promulgated in time to affect the initial permits for the proposed facility, and the city appeared to interpret this to mean that information on toxic air emissions was generally not required to be provided or considered under CEQA, except to the extent that state and local regulations required air toxics-related compliance.

Air toxics had been recently addressed at the state level in the form of legislation known as the Air Toxics Hot Spots Information and Assessment Act of 1989 (AB 2588). Primarily information oriented, AB 2588 requires certain manufacturing facilities to prepare a health risk assessment (HRA), based on estimated air emissions of more than two hundred listed substances. Although an HRA was prepared for emissions from the proposed incinerator, it was not prepared under AB 2588, but rather under the authority of California Health and Safety Code Section 42315, which requires the preparation of an HRA for "any project which burns municipal waste." Thus, it did not consider potential hazardous air emissions from any project sources other than the incinerator, nor did it explicitly take into account the more than two hundred substances identified under AB 2588 or provide estimates of their release. Furthermore, the population cancer risk calculated in the HRA was based on a city population of 28,000, although the population of West Sacramento was projected to be at least 40,000 by the time the project was completed. It also failed to consider potential population exposures in the adjacent city of Sacramento, although the project site was only a few miles from the state capitol building in downtown Sacramento.

The reasoning behind the decision to consider only incinerator sources of air toxics derived from the segmention of the project not only by construction schedule, but also by ownership. This segmentation scheme became the centerpiece of the developers' strategy for securing environmental permits. As the EIR stated:

> Phase Ia [the incinerator] would be owned and operated by one company (Hipp Engineering, Ltd.), while Phases I, II and III would be owned and operated by [MacMillan Bloedel and Haindl]. For permitting purposes, the Phase Ia facility would be considered separately from the Phase I, II and III facilities.[10]

(Hipp Engineering was the MacMillan Bloedel/Haindl project engineer and also the designer of the Newstech mill in British Columbia.) Thus, the reasoning went, the other phases would not burn municipal waste, so they would not require HRAs. The on-paper isolation of Phase Ia by ownership was widely condemned (by critics including the California Energy Commission, the Yolo/Solano Air Pollution Control District, and the Sacramento Metropolitan Air Quality Management District) as a transparent device to evade regulation, but this condemnation failed to move the city council.

Other respondents pressed for the comprehensive disclosure of estimated air toxics releases. Both the Sacramento Valley Toxics Campaign and a research group from the University of California at Los Angeles, requested chemical-specific estimates modeled upon the approach required by both AB 2588 and the federal Emergency Planning and Community Right-to-Know Act (i.e., TRI reports). As the city continued to insist, however, the facility would not be *required* to comply with these reporting requirements until *after* it was built. Ultimately, no comprehensive estimation of the identities or amounts of toxic substances to be released to the air or other environmental media was ever provided.

The segmentation by project phase and ownership proved highly useful to project proponents in other areas as well. With respect to criteria air pollutant regulations (affecting emissions of carbon monoxide, reactive organic gases, nitrogen oxides, sulfur oxides, suspended particulate matter, and lead), the project developers proposed a similarly evasive compliance strategy.[11] Phases I and Ia, with air pollution permit applications already on file under two different company names, would be permitted first and separately; relatively soon thereafter Phase II would be permitted as an expansion of Phase I. This essentially allowed "the project," through Phase II, to be artificially redefined as two separate projects ("Phase Ia" and "Phases I and II"), and each of these individually to slip under the action levels defined for new sources of criteria air pollutants. These action levels triggered a strict set of requirements: the installation of best available control technology (BACT) and the purchase of "emissions offset credits" so that no net increase in the triggering pollutant would occur in the air basin.[12]

Had the projected emissions of Phases I and Ia alone been aggregated, the BACT and emissions offset action levels would have been triggered for at least two of the pollutants to which they applied (nitrogen oxides and reactive organic gases). The regional market for emissions offsets was, however, extremely limited. In particular, NOx offsets were currently unavailable, although earlier in the year the Sacramento Municipal Utility District (SMUD) had announced an agreement to purchase 260 tons per year worth of NOx offsets from Pacific Gas and Electric at a cost of $7.28 million. Assuming NOx offsets became available, and assuming the same unit cost that SMUD had paid, the purchase of offsets for the recycling project's NOx emissions alone would have been more than $2.2 million for Phases I, Ia and II. If, as seemed likely, the ratio between new emissions and offsets was required to be greater than 1:1, the cost would exceed $3 million. The unavailability of NOx offsets could have created temporary complications for the project.[13]

Ultimately at issue was the daily release of several thousand pounds of regulated air pollutants in an air basin already in noncompliance with air quality standards for three of these pollutants. Nevertheless, the EIR found a "less-than-significant" air quality impact due to the apparent ability of the (segmented) project(s) to comply with regulations. Few others shared this view, and the certainty of regulatory approval was by no means guaranteed. As the deputy director of the California Energy Commission observed:

> The phrase "less than significant impact" as used here has no relevance to the physical environment and the changes in air quality that may result from this project . . . This is especially critical in criteria pollutant nonattainment areas, where ambient standards are already being exceeded and therefore significant public health impacts are already occurring.[14]

Also included among those not persuaded by the approach advanced by the developers and the city was the senior staff counsel to the California Air Resources Board who wrote in a letter to the City's chief planner, "[B]y not proposing to obtain offsets for Phases I and Ia of the project's emissions, the project is in violation of both the letter and spirit of the Air Resources Board transport mitigation regulations." She added:

> . . . [T]he concept of a wastepaper recycling facility is environmentally laudable. However to sully the project through economic blackmail and regulatory shenanigans does us all a disservice. At least in the narrow area of air quality impacts, the project proponent should come clean and obtain offsets, as required by law, logic, and common decency toward the public.[15]

If the approach to air pollution was complex and evasive, the approach to water pollution was simple and evasive. Improving on the segmented project approach, the city concluded that the impacts of discharging 7.8 million gallons per day of treated project effluent into the Sacramento River would be the subject of a separate, subsequent EIR, addressing the environmental impacts of the proposed municipal wastewater treatment plant (WWTP). As the revised draft EIR for the newsprint facility observed, "The precise extent of the beneficial effects of the city's [future WWTP] on MacMillan Bloedel/Haindl wastewater prior to discharge to the Sacramento River is unknown at this time."[16]

Although the WWTP was still in the conceptual planning stages, the EIR for the recycling project nevertheless provided some preliminary estimates of its effectiveness that suggested certain concentrations of pollutants in the treated wastewater. The estimates were disturbing enough to set off alarms in offices throughout the state. The Regional Water Quality Control Board raised major concerns, including the lack of assimilative capacity in the river for estimated concentrations of heavy metals, suspended solids, phosphates and salts; it also identified numerous errors in the representation of water quality standards and criticized the absence of any estimates concerning how much additional water might need to be released from upstream reservoirs in order to dilute the plant wastewater to Delta water quality standards. The State Department of Water Resources criticized the absence of data on discharges of trihalomethane (THM) precursors and dioxins.[17] (THMs are suspected human carcinogens formed by the chlorine-based disinfection of water containing organic precursors, measured by total organic carbon [TOC] and bromide.)

Perhaps the loudest chorus of protest over both potential discharges and the segmentation of the EIR process came from state water contractors, including, most significantly, the powerful Metropolitan Water District (MWD) in Los Angeles. These water agencies contract with the state for the annual delivery of more than 3.6 million acre-feet of water from the State Water Project (SWP) and serve more than twenty million of the state's residents, of which the MWD alone serves some fifteen million. The SWP pumps water out of the Delta into an aqueduct that extends four hundred miles south to the Los Angeles basin. The withdrawal of water for the SWP is limited by the need to maintain adequate levels of freshwater in the Delta, which is subject to excessive salinity intrusions from downstream and high concentrations of other pollutants if not adequately flushed. The prospect of having to release more water from upstream reservoirs to dilute discharges from the WWTP—water that would be subtracted from that available to the water contractors—was not something they took lightly. Tentative estimates calculated by the State Department of Water Resources had suggested that as much as 200,000 acre-feet of water per year might be at issue.[18]

Equally troubling to the water agencies was the potential impact of increased THM precursors. Because SWP water supplies urban water users, water agencies must meet the EPA's standards for THMs in drinking water, standards that were expected to become substantially more restrictive in the future. As MWD pointed out, "Any increase in the [TOC] or bromide in the State Water Project could conceivably force the installation of very expensive advanced treatment technology [by the water agencies] in order to meet even current requirements, much less more stringent future standards."[19]

Despite being put on alert by essentially every player in California who had either direct authority or significant influence over the subject of Delta water supply and quality, the city council nevertheless stood fast in its decision to segment the EIR process and in September approved both the final EIR and the conditional use permit for the recycling facility without considering the Delta water impacts. Even without taking the water issues into consideration the city found that the project would have a "significant detrimental impact" on the environment; yet, it concluded that "overriding considerations" warranted approval of the project permit.

The primary overriding consideration was a much advertised but never substantiated $4 million in annual redevelopment tax increment revenues to the city. This estimate was based on the debatable assumption that the project equipment would not be depreciated; it also incorporated revenues from other business activities projected to be indirectly generated by the facility and was derived from an economic impact assessment conducted by a consulting firm hired by MacMillan Bloedel—the chief project beneficiary. The city's Economic Development Advisory Commission held public hearings based on this report and appeared, along with the city council, to be motivated by an interest in securing some $3.7 million of the hoped-for tax benefit in the form of redevelopment funds to be applied toward the development of infrastructure in the Southport Industrial Park (SIP). As the city noted in its Statement of Overriding Considerations, "The proposed project would provide an anchor for future development of the SIP [and] would result in sufficient development to begin financing public infrastructure in the SIP."[20]

It was, in short, a sort of municipal version of a pyramid scheme based on growth. The cost to the city of building the required WWTP—which, it appeared increasingly likely, might have to include a desalination plant, and granulated activated carbon treatment for TOC—was never conclusively documented, nor was an independent assessment of the costs of other infrastructure development (including bridges over the adjacent deep water shipping channel, sewer lines, etc.) ever made public. When community residents asked for the city's economic assessment they were given a summary of the MacMillan Bloedel-commissioned report.

The report estimated the WWTP cost at $10 million, and other infrastructure development costs to the city at approximately $3.5 million. It eventually emerged that the city had commissioned its own economic assessment of the project, but when members of QUAD sought to obtain a copy from the city they were informed it could not be made available because it contained subject matter pertaining to "confidential negotiations." The negotiations apparently concerned what the city would offer the project developers as financial inducements, including potential use fee waivers for city services, advantageous property tax treatment, and the construction of the WWTP. In one meeting of the Economic Development Advisory Commission, the Commission chair bluntly asked, referring to the negotiations with the developers, "How much are they asking for?" After some hesitation, the Director of Redevelopment responded, "Fifty million dollars."[21]

Although never provided with an independent appraisal of the financial impacts of the project or of the costs of providing what would clearly have to be a state-of-the-art WWTP, some city residents began to feel that the financial benefits looked rather dubious. Real financial benefits that could be applied to schools or public amenities seemed to depend on further (probably also controversial) industrial development that would come, if at all, some time in the future, after tax revenue from the recycling facility had been channeled into infrastructure development to make such future growth possible. In any event, the public financial costs and benefits of the project were not specifically germane to the environmental review under CEQA and therefore were not presented as part of the EIR process. As the city observed, "CEQA does not require lead agencies to analyze the economic impacts of proposed projects. Rather EIRs focus on potentially significant environmental effects. . . . Economic or social effects are not considered to be significant environmental effects within the meaning of CEQA."[22]

The only real question remaining by the time the city certified the final EIR was, Who would sue first? Although the project appeared to be headed for regulatory problems somewhere down the line, opponents were aware that each hurdle overcome by the developers would bring the project closer to construction. When the water agencies appeared to be reserving their legal options for the EIR on the wastewater treatment plant, the Sacramento Valley Toxics Campaign and various individuals filed suit in October 1992. They alleged that the "piecemeal, segmented environmental analysis" had resulted in a "bad faith circumvention and violation of CEQA" and that the project would have a "substantial, immediate, and long-term detrimental impact upon the people, the physical environment and the economies of Yolo County and Sacramento County and the other Sacramento Delta counties."[23]

This certainly did not sound like what the recycling rhetoric of Earth Day 1990 had promised, or like the high hopes reflected in California's AB 939 or even in RCRA. In fact, the various benefits of recycling had been all but forgotten in the struggle to respond to the immediate threat to the community posed by a huge new industrial development on a very fast track. Among the environmental groups involved, only Californians Against Waste, a Sacramento-based organization that had been active on solid waste and recycling issues for more than a decade (and had sought to position itself as the mediator between MacMillan Bloedel and the regional environmental community), continued to express support for the project and to emphasize the "net" benefits.

Also lost in the process was the fact that many members of the community would have been quite happy to see a smaller recycling project in the same location: the main problem with this facility was that it was vastly oversized. Under CEQA the EIR was required to present alternatives to the project, but only one of the five alternatives discussed had proposed a smaller version of the facility—to be accomplished by eliminating Phase III. This alternative was, however, summarily dismissed as "infeasible" due to "economy of scale" considerations.[24] Unfortunately, the developers' economies of scale had a classically inverse relationship to the quality of the regional environment, and the financial feasibility of the project could not be assessed either as proposed or under alternate scenarios. One of the few references to the expected profit margin that emerged in public documents was made by a MacMillan Bloedel executive who claimed, while addressing the subject of purchasing NOx offsets, that "given the already marginal financial return on the project (about 6 percent) this additional cost would further jeopardize the likelihood of this project proceeding."[25] This figure was echoed in a table of cost indexes that MacMillan Bloedel submitted as part of its commentary on the revised draft EIR. However, in this document the company estimated a 6 percent return on Phase I and a 22 percent return on Phase II, for a net return on both phases of 15 percent.[26] Based on the several smaller mills being proposed or developed around the country at the same time, it seems safe to conclude that the developer's claim that the project was infeasible at smaller scales was rather exaggerated.

The Distribution of Benefits

Overall, the city seemed immune to an appreciation of the difference between local impacts and regional or state benefits and to the idea that it actually mattered *where* the various environmental and economic costs and benefits would be distributed. Despite the controversy over diminished local landfill capacity, for example, the city wrote "viewed on a State-wide, rather than *parochial* level, the MB/H plant would increase the useful life of the State's landfills" (emphasis added).[27] It was not surprising that the project was widely understood to have the support of the governor of California because the city council seemed to be viewing the project from his perspective. When EIR respondents pressed for further consideration of a scaled-back facility, the city observed, "If

California is to attain its waste newsprint recycling goals, any downsizing of the MB/H plant would have to be offset by locating additional recycling plants elsewhere in the state or by exporting. Each additional site would be accompanied by environmental consequences of its own."[28]

It was not, however, simply a matter of relocating one-half or one-third or one-fifth of the same environmental impact in multiple locations through a series of smaller projects. In fact, unique opportunities and benefits could be realized only at smaller scales: a wider variety of options, more community control, far more flexibility for experimentation. For example, a pilot project testing the effects of careful monitoring of wastepaper supplies to eliminate contaminants that influence the toxicity of the process sludge, and thus influence its potential use as a composting or soil amendment, would look more far more palatable if the fallback position in the event of failure were to landfill 200 tons of sludge per day, instead of 1,000 tons per day.

Similarly, one city resident had suggested the possibility of constructing a wetlands system for treating project effluent, citing a number of such projects in operation (two of the largest and most successful of which were located within a few hours drive of the city). It was a provocative idea, particularly in light of the vast wetlands that had once characterized the area and of increasing state interest in conserving and restoring wetlands. The city agreed in principal, although it rejected the approach based on what it claimed were problems that could be anticipated due to the expected high salt concentrations and high temperature of the effluent.[29] More important, however, was that the biologically complex and comparatively experimental nature of a constructed wetlands approach was likely not compatible with the fast-track, large-scale, growth-inducing nature of the MacMillan Bloedel/Haindl project.

There was also an argument to be made for proceeding at a smaller scale and forestalling a huge, irrevocable capital commitment to a single technology in light of the accelerated pace of innovation in paper recycling technology that had finally begun to occur. The domestic technical literature on paper recycling, stagnant throughout most of the 1980s, had rapidly expanded by 1992. Indeed, the prospects for closed-loop deinking (the elimination of effluent through treatment and reuse of process water) had already emerged as a serious objective.

Although the Sacramento area would eventually gain somewhere between 350 and 450 permanent jobs (the estimates fluctuated) when the facility was completed, the city of West Sacramento itself was not necessarily getting a great deal in terms of employment development in light of the environmental and financial burdens it would be assuming. The new capital invested by the developers would amount to approximately $4.5 million per job created. Assuming that the costs of the city's commitment to build the WWTP (with an estimated 80 percent of initial capacity dedicated to the mill) and other infrastructure would be at least $20 million, the city could be expected to finance each job to the tune of between about $50,000 to $80,000 (the higher figure if Phase III were never built), not including interest on the debt. Clearly, the city's costs might run significantly higher. On the other hand, the project was strongly supported by the Sacramento-Sierra Building and Construction Trades Council, which had backed the campaigns of several city council members. A representative who testified in a public hearing on the project (packed with Construction Trades members) stated that "with work like this, our members will be able to participate in the West Sacramento dream,"[30] most likely a dream of future growth supported by project-related infrastructure development.

Both proponents and opponents also requested more information on the overall net benefits to the environment, particularly in terms of forest conservation and pollution prevention. The city responded first by pointing out that "CEQA does not require lead agencies to analyze 'beneficial impacts' in an EIR" and then by providing yet another recitation of the statewide solid waste benefits.[31] In reference to claims in the EIR that the project would reduce air pollution, conserve energy, and save trees, QUAD asked, "Where will the air be cleaner," "Which facilities will be shut down or reduced in capacity," and "Which forests will be spared?" The city responded with the conventional wisdom that "every ton of recycled paper would displace the production of a ton of virgin paper," observing, however, that "it is impossible to know which forest(s) would be preserved."[32]

The dialogue had become highly antagonistic by the time of the final EIR and the QUAD questions were likely viewed by project proponents as a last-ditch effort to trip up the process. At another level, however, the questions reflected critically relevant issues that should have been considered at some point, even though the EIR process did not seem to be the right forum. Given the controversy raging over the destruction of western forests, including the liquidation of British Columbia's ancient forests in which MacMillan Bloedel was centrally involved, it was not unreasonable to wonder where, if anywhere, the project would promote a cutback in the production of virgin paper. For its part, MacMillan Bloedel was making no such promises and had in fact pursued and acquired cutting rights to the pristine Clayoquot Sound forest on Vancouver Island during the time that the controversy over the West Sacramento project was on-going.[33]

Nor were there any promises to hold back per capita consumption of paper from further growth in the state or to accompany the recycling project with source reduction planning. Indeed, it was a replay of the fight that had sharply escalated in the mid-1980s (and was still being played out) over the hierarchy of waste management options in which recycling was rhetorically preferred over incineration, but incineration received the lion's share of federal and state support. In this case, source reduction was being ignored in the rush to dramatically expand recycling capacity, an omission only exacerbated by the range of hollow environmental claims being made around the project.

Developing a Regional Perspective

When the point of view is shifted from "should the MacMillan Bloedel/Haindl mill be approved" to "how should paper recycling be developed in California" (or in any other state or region for that matter), ideas and alternatives build very rapidly and begin to spin off across a web of interconnected industrial and ecological relationships. The two points of view reflect the difference between a reactive posture and a proactive posture. The EIR process in this case, manipulated by powerful corporate interests, with the help of a developer-backed city council that saw its responsibilities to the community framed by the developers' arguments, was barely adequate to address the first question and hopelessly unsuitable for dealing with the second. Although the state had, through passage of AB 939, finally begun to assume responsibility for promoting paper recycling as an alternative to landfilling and had tentatively addressed the problem of finding a market for ONP through minimum content standards in newsprint, it had failed to go any further. There continued to be a vacuum in relation to the state's capacity to establish a larger environmental and economic policy framework in which to consider either the project itself or the more general subject of expanding wastepaper-based production capacity in the state. Indeed, the most likely candidate for coordinating state policy on such matters, the California Integrated Waste Management Board, continues to be silent on the subject of developing recycled paper production capacity.[34]

Among the lessons and policy implications that arise from the West Sacramento case, several issues stand out. The first, regarding the EIR process itself, is that although the documents ultimately provided a wealth of essential information on the facility (it is disconcerting to imagine either a past or a future when an industrial development of this scale could move forward without detailed public disclosure and review), the project had clear overtones of a "done deal" by the time the CEQA requirements were triggered. The alternatives proposed in the EIR, for example, may or may not have met the letter of the law, but they were widely viewed as insincere and inconsequential. Any contention that they represented a reasonable assessment of the range of viable alternatives is not only extraordinarily shortsighted, but also saddening in light of the real alternatives that exist, of which many were identified by the community.

Indeed, a core group of individuals who had been involved in the process did emerge with the capability of characterizing and seeking development of a paper mill that would be appropriate in the local context, although not until they were presented with what amounted to a worst-case scenario for recycled paper manufacturing. Out of the need to react came a process of imaginative thinking, research, networking, and speculation around subjects as diverse as wetlands, green space, integration of recycling and agricultural residue use, forest conservation and potential linkages with northern communities in the state, small-scale pulping technologies, and alternative models for financing and job creation. This powerful local resource, essential in establishing facilities not only with community support but drawing upon community resources, was never considered by either the developer or local public officials, nor did the city have an economic development process that was organized to generate rather than react to proposals.

In addition to the fundamentally reactive nature of the EIR process, which nevertheless stood in as the primary vehicle for democratic community planning, a second problem was its artificial disconnection from economic impacts. The anticipated financial benefits to the city—which constituted the overriding consideration for approval—were highly questionable and never actually substantiated. That the only publicly available analysis of economic impacts was paid for by the project developers obviously rendered it unreliable if not useless in light of the evasive and irresponsible approaches they had favored in their attempts to secure environmental permits.

Overall, the conflict highlighted the inadequacies of the state's highly generalized, solid waste management-driven "recycling market development" approach to finding outlets for the increasing supplies of recovered wastepaper. It also highlighted the importance of considering local outcomes of sweeping state policy objectives. There was—and is—an absence of any alternative planning and review framework for informing this kind of industrial development with more deep-rooted social and environmental criteria. The EIR process, however important, was nevertheless a poor match for the powerful array of forces behind this proposal: a David in the face of Goliath (or Godzilla, as the project came to be identified in the community). More active and democratic planning processes are certainly to be recommended in general, but there will always be a gap in the ability of communities to leverage their relationships with companies and industries that are fundamentally national and international in scope and structure. More than the project itself, what really needed to be fixed was the calculus that gave rise to it.

A public sector that passively abdicates responsibility for considering the industrial response to a major new materials stream it has set in motion should not be surprised if it is offered very large-scale, vertically integrated, capital-intensive, monopolistic projects that view public health and the environment in terms of a series of permitting problems to be overcome by whatever means are available and that manage to turn one of the most important environmental and economic development opportunities in recent memory on its head. This same public sector should not be surprised when they are opposed.

The MacMillan Bloedel project was ultimately permitted by air quality regulators (after significant concessions by the developers) and by the city government. The lawsuit was eventually settled when the developers agreed not to build Phase III, and the plaintiffs agreed not to talk about the project in public. The water agencies and regulators have not had to take on the wastewater issues, as changes in the newsprint market in the United States altered the economics of the mill and thus of the wastewater treatment plant. By 1996, neither project had moved forward and the mill proposal had apparently been abandoned. California continues to generate and recover immense quantities of wastepaper each year, the majority of which continue to be landfilled or exported.

Neither beneficial wastepaper recycling development nor long-range transformation of the paper industry will likely be realized through a fragmented, business-as-usual approach to economic development, consumer education, resource management, or pollution control and waste management, or through traditional forms of policy, advocacy, and regulation narrowly connected to those issue areas. As argued from other contexts in preceding chapters, a new approach must begin by understanding and addressing the choices that will influence the future of the pulp and paper sector as a whole. One can claim that the existing paper industry is too big to be transformed from the top down, and that the opportunities for effective federal leadership on the issue are presently nonexistent. Yet these may be blessings in disguise. If a sustainable paper industry must be built upon the use of regionally diverse fiber resources, there are no one-size-fits-all solutions. At the same time, critical opportunities for change already exist around the potential to influence the development, conversion, or expansion of individual mills, to make regionally-unique fiber resources available, and to influence and educate local markets in terms of both paper acquisition, and paper use efficiency. These opportunities lend themselves to coordinated efforts focused at local, regional and state levels.

Although California is the largest consumer of paper in North America (consuming more than twice as much paper as all of Canada) and is awash with alternative fiber resources including agricultural residues, it is almost totally dependent on pulp and paper imports. Due to the concentration of the industry, many states will find themselves in the same position, however much they may differ on all other dimensions. California is not alone in being a compelling candidate for the development of alternative fiber-based pulp and paper production capacity geared toward state markets, and thus for the reduction of its dependence on wood fiber and imports. There are an increasing number of models for mills that can be commercialized at small or moderate scales using clean, energy and water efficient pulping and deinking technologies. The momentum around nonwood opportunities is building and can be enhanced.

The interrelationship between production and consumption can also be considered: the development of socially and environmentally informed regional production capacity could proceed naturally in conjunction with efforts to identify the characteristics of a sustainable paper market. Ironically, the present gap in pulp and paper production capacity in the state—the absence of a large established industry that might be threatened—may well constitute an opportunity for seriously addressing both environmentally informed paper acquisition and highly efficient paper use. Taking a national perspective, some have argued that "source reduction strategies need to be sensitive to concerns about loss of business and jobs in the affected industries," but few have considered that in many regions there could be an inverse correlation when use reduction efforts are linked, as they naturally are, with environmentally informed procurement efforts.[35] Educational and assistance efforts focused on encouraging a preference for declining virgin wood fiber and chlorine content in paper, as well as for decreased consumption relative to function and service, could be pursued without serious challenge to an established industry in California, in such a way that they could help support and cultivate an emerging state industry.

A new environmental discourse—one that includes economic and social content as equal partners in the development of environmental policy and strategy—has a natural geographic and political affinity with the longstanding economic and industrial development missions of states and local governments. The potential of such a discourse is significant, but what continue to be missing are the ability to escape the narrow policy boxes in which the relevant issues have been stranded and a commitment to find ways to examine more long-range industrial visions from within the framework of regional opportunity. This approach at least promises to be more interesting than the standard alternative. As West Sacramento's redevelopment director tiredly observed, "We're getting used to the fact that every time we bring in a new project someone challenges it."[36]

6

Prospects and Challenges

The most daunting task facing those working to define the policies, strategies, and tools that can be used in the march toward more sustainable systems of production and consumption is the need to confront modern precepts of public planning, economics, and economic development. Many within the extended environmental community, including some economists, have come to understand environmental problems as fundamentally embedded in core tenets of received economic wisdom. This environmental critique extends equally to premises that have guided both capitalist and socialist market economies as well as to the unmistakable legacy of nonmarket industrial economies. Targeting the worsening disconnection of economic theory from observed reality, Herman Daly and John Cobb have argued:

> Whenever the abstracted-from elements of reality become too insistently evident in our experience, their existence is admitted by the category "externality." Externalities are ad hoc corrections introduced as needed to save appearances. . . . Externalities do represent a recognition of neglected aspects of concrete experience, but in such a way as to minimize restructuring of the basic theory. As long as externalities involve minor details, this is perhaps a reasonable procedure. But when vital issues (e.g., the capacity of the earth to support life) have to be classed as externalities, it is time to restructure basic concepts.[1]

The ascendance of the GDP, which leaves the accounting of environmental degradation and associated human impacts either off the books or reflected as a plus, as the overriding indicator for national achievement is perhaps the most obvious and pernicious modern symptom of economic indifference to the environment. The wholesale national pursuit of industrial economic competitiveness in isolation from any thorough examination of the environmental and social "competitiveness" of the ascendant industries follows. The substitution of a lonely materialism and media saturation for community life and personal development is yet another symptom.

It goes without saying that resolving these deep contradictions in more than a patchwork way will be a task that will continue to engage thinkers and actors from many different perspectives for generations to come. Yet, the magnitude of the task of thoroughly revisiting established economic principles (or even the past half century of U.S. economic policies) from an ecological perspective and of rebuilding the mediating institutions of the political economy in this light is equalled by the urgent need to act effectively to reverse the rate of growth in environmental degradation. For the foreseeable future environmentalists will continue to be working to effect substantive progress within the inhospitable context of an increasingly global economic system that is, at its roots, fundamentally hostile to the environment. At the same time, they and others will continue to seek new definitions and measures of development, growth, progress and productivity grounded in what André Gorz has called "ecological realism" and in its partner social justice.[2]

Whether, as an environmentalist, one does or does not subscribe to a belief that dominant economic premises require more than tinkering as a fix on the road to sustainability, there is plenty of work to go around in the short term. Oddly enough, however, those most committed to a more fundamental critique may well find themselves out of necessity grasping many tools and measures that seem most representative of the economic order they criticize. There will, however, be a sometimes subtle and sometimes overt difference in how they are applied and in who applies them. A practical outcome of this necessity is that many of the new formats, materials, and technologies important to environmentally informed restructuring in the paper industry can and will be commercialized even though they will continue to operate on an uneven playing field, and their nonmonetary benefits will continue to go largely unheralded and unaccounted for in the mainstream schemas.

This chapter is oriented toward the immediate task of exploring practical opportunities that exist even within the contradictions of current economic frameworks and economic globalism. Fortunately, important concepts and tools are available to assist in this difficult undertaking.

Many can be borrowed, with adaptation, from disciplines and schools of thought that are otherwise often blind to the environment. The concepts and tools examined here include: (1) sectoral analysis and policy, (2) regional planning and development, and (3) information-oriented regulation. The latter two subjects briefly emerged in the preceding discussion of the case study of West Sacramento and in considering the stimulus provided by the Toxics Release Inventory, but sectoral analysis has been central. The following sections provide some elaboration of these subjects in terms of how they may guide and assist practical near-term efforts focused on developing the foundations of a sustainable system of paper production and consumption. The industry itself continues to have an enormous opportunity to further these goals; however, the following discussion is primarily oriented toward informing efforts that can be undertaken in the public sector and expressed in both governmental and nongovernmental forms.

Sectoral Analysis and Policy

Sectoral policy proceeds from sectoral analysis, and efforts in the latter (often leading to recommendations for the former) have long been pursued within the framework of macroeconomic analysis and policy. Many scholarly discourses on the evolution, structure, and location of different industries arise principally from the disciplines of economics, economic geography, and regional planning. Yet, with painfully rare exceptions, when one reads these studies from an informed environmental perspective, it is like stepping into a parallel universe in which environmental reality has been wiped from the slate. Learned treatises on the steel industry or the petrochemical industry or the miracle of Silicon Valley and on the lessons we can take from their examples are construed and elaborated without a whisper of the grim tolls they have taken on human health and the environment.[3] The substantive incorporation of these "external" realities rarely occurs. One occasionally obtains advice such as: "plants should incorporate pollution control technology." Natural resources (raw materials) must of course be mentioned, but they are simply "plentiful" or "used up": cheap or expensive. The massive ecological trauma that was certainly involved in moving from one to the other is nowhere to be found. The wonders of twentieth-century agricultural productivity are currently extolled as an unconditional model for the forest products industries, as if Rachel Carson had never lived and as if much of modern environmental effort were not focused precisely on undoing the social and ecological consequences of large-scale chemical-intensive agribusiness.

The preceding chapters have been focused on sectoral analysis primarily from an environmental perspective—more accurately, from multiple, traditionally isolated environmental perspectives: resource policy, industrial pollution control, and solid waste management. This approach can alternately be viewed as industry-centered environmental analysis or as environmentally grounded sectoral analysis. The analysis has suggested that the integration of these multiple environmental perspectives produces the best guidelines for future development of the industry. It has further suggested that these environmental guidelines are innately more attuned to issues of employment and community life conventionally understood within a social justice framework. The strongest link between the two areas can be seen in the need for greater material and regional diversification and in the attendant need for smaller production scales.

As a way to approach complex, multidimensional environmental issues, the value of sectoral analysis has been insufficiently appreciated to date. The approach is not simply about identifying the range of environmental problems associated with a sector, understanding the ways they interact, and suggesting a more integrated environmental regulatory strategy. This approach is begun, although not comprehensively, with the EPA's Common Sense Initiative. The potential of an environmentally grounded sectoral analysis is equally about understanding environmental problems in terms of how they are embedded in the forces that define and influence the larger industrial system.

For example, the linkages between the issues of chlorine use and wood use were examined in chapter 3. One may choose ECF or TCF strategies that are, by default, oriented toward reinforcing wood use (involving large-scale, rapid reinvestment in woodpulp mills), or one may pursue strategies emphasizing alternative chemical pulping technologies and alternative fiber sources. In either case, prospects for effective intervention, whatever form it takes, are substantially enhanced if one also has some grasp of the ways in which pulp production is integrated with timber and other forest products production, the locational and scale barriers that attach to alternative materials use, the R&D capacity of the industry, the central technological themes, and the strong, longstanding economic linkages between the pulp and basic chemicals sectors. The environmental problems are related by the sector's internal structural characteristics, which in turn are heavily mediated by its relationship with other sectors.

Although these networks of structural interactions are understood and followed in economic analysis (at least in terms of capital flows, if not always material flows), and their more complex biological analogues are the basis of the science of ecology, they are poorly reflected in most environmental policy analyses. As suggested throughout, environmental analysis (and advocacy) often devolves into traditional, isolated policy areas or into a focus on individual companies, facilities, resource supply regions, or products. Product life-cycle analysis, advanced as *the* systems methodology for environmental analysis, has exacerbated this tendency.[4] Environmentally grounded sectoral analysis, however, offers an alternative approach to systemic environmental policy analysis; one which is necessarily inclusive of issues of employment, place, and change. It also offers the practical advantage of isolating a meaningful subset of an otherwise unapproachably complex economic system: environmentally grounded sectoral analysis is both complex and containable. In this respect, it may help bridge the gaps that lie between real-world problems—decisions now before us—and long-range goals of sustainability. The attempt to understand and effect fundamental progress on a sustainable paper industry, particularly at a regional level as discussed below, is both possible and worthwhile. The complexities of the system can be honored, grasped, and translated into action more easily within this framework than when operating within the abstractions of a sustainable economy or the limitations of conventional, piecemeal policy frameworks.

Sectoral analysis, however, is different from sectoral policy, and it is necessary to contain the latter at the outset because it is a short leap to industrial policy. Indeed, the terms industrial policy and sectoral policy are somewhat interchangeable (industrial policy tends to devolve to sectoral policy), and we need to look before we leap.

The debate over industrial policy arose in the 1970s in response to economic dislocations that began to appear at the end of the 1960s after two decades of rapid postwar industrial growth and expansion. Annemieke Roobeek differentiates the national governmental responses that developed within the industrialized countries through the 1970s and into the 1980s into *defensive industrial policies* and *offensive technology policies.*[5] This difference is also understood in terms of backing the losers (or at least softening their decline) versus backing the winners. Early industrial policy debates tended to focus on faltering basic industries such as steel, whereas technology policy debates have tended to focus on high-technology "sunrise" industries. The general goal is to achieve economic prosperity and social stabilization through government-assisted industrial competitiveness, especially during periods of restructuring and change. The historical debate regarding the role of government intervention in industrial development is traced to two main lines of reasoning set out in the eighteenth century. One is the laissez faire doctrine set forth in Adam Smith's *The Wealth of Nations;* the other is the prointerventionist argument laid out in Alexander Hamilton's *Report on Manufacturers.*[6]

The actual tools and mechanisms that could be deployed in operationalizing an industrial policy by the 1970s cut a wide swath. Generally, an industrial policy implied some need for the more or less coordinated marshalling of various existing government powers around a focal point defined by a particular industry or set of industries, one or more core technologies or companies, or some combination thereof for the broad purposes of serving the national interest. Thus, such policies were both distinguished from and heavily overlapped with more generalized economic and social policies. The set of powers that could be called upon, although tending to be most clearly highlighted by targeted public funding of research and development programs in the case of technology policy, were ultimately no less than the full set of powers held by the government.

The strongest argument advanced for industrial policy by the early 1980s was a simple one. It was based on the assertion that the U.S. government had long been so deeply engaged in industrial policy that it amounted to one of its main activities. A 1983 article in *Business Week* asserted, "The United States has been in the industrial policy game at least since 1643, when Massachusetts granted a new smelting company exclusive iron-producing privileges for twenty-one years to encourage the industry."[7] However, almost endless and much broader examples could be found. They included government support of what can alternately be viewed as public or industrial infrastructure in the form of building transportation systems, dams, and even university systems, not to mention the bankrolling of synthetic rubber, semiconductors, nuclear power, agriculture, and, of course, the military-industrial complex as a whole. Industrial policy could also be seen in more or less active modes: imposing selective trade tariffs or pouring billions of dollars into loans and loan guarantees to support risky export/import ventures. Getting "the government off business's back" could be viewed as a passive form of industrial policy, and antitrust regulation as a more active form. What was then (in the early 1980s) a $750 million a year tax break to the timber industry was appropriately identified as yet another form of industrial policy. Ultimately, of course, there is very little the government does that does not have an effect on the economy (hence, political economy) and thus on industry. The conclusion reached by this expansive view of industrial policy was that the United States should come clean, "drop its guise of economic innocence," and start rationalizing its industrial policy, aiming for greater efficiency and competitiveness.[8]

The problem, of course, was who would do the rationalizing. As Alan Greenspan, then an economic advisor to Ronald Reagan (now chairman of the Federal Reserve Board), argued, "industrial policy is a mechanism by which the politically powerful get their hands in the till." Unfortunately, this appeared to be true regardless of whether industrial policy continued to be covert, or became overt. *The Nation* agreed with Greenspan on this point, arguing against industrial policy proposals framed largely by centrist democrats because "they would transfer resources to the wealthiest and the largest corporations."[9] At least, they would do so in a worse way than was already occurring. A central problem with coming clean about industrial policy was that many specific proposals reflected further circumvention not only of the market process, but also of the democratic process, tending to feature various centralized planning boards or banks consisting of a mix of labor, management, and government. The science of muddling through within existing democratic and market processes was seen by many as preferable to the abuses that would certainly arise from elitist rationalism and planning, no matter how much the rhetoric seemed to reflect a sincere concern for the public welfare. The issues of industrial policy inevitably raised the question of *whose* industrial policy.

Predictably, one had to look long and hard to find an environmental perspective on any of this. One came in the form of the 1984 report from the Project on Industrial Policy and the Environment written by representatives of ten environmental groups (referenced in chapter 1). Most interesting, however, was not the report itself, a generally bland endorsement of both "new" industrial policy and status quo economic policy, but the impassioned dissent written by Richard Grossman of Environmentalists for Full Employment. This dissent was joined by the representatives from Public Citizen's Critical Mass Energy Project, Environmental Action, and the Urban Environmental Conference. The nondissenters included the Environmental Defense Fund, Natural Resources Defense Council, Sierra Club, National Audubon Society, and the Wilderness Society, reflecting a classic mainstream/alternative split in positions.[10]

The dissenters flatly rejected the endorsement of international competitiveness and export boosting as the appropriate orientation for industrial policy, arguing that the real basis for the discussion was the fact that "people and communities are challenging existing power structures which have not been responding to their needs." As Grossman wrote, "people are breaking through the traditional limitations of discourse on economic issues and seeking ways to redefine their 'problems.'. . . They are talking jobs and income policies, not helping multinationals sell their crap all over the world with the hope that some jobs will show up here." The dissenters also highlighted the failure to consider recent economic history and root causes, including the worsening expropriation of the public role in decision making by growing corporate control. Overall, the dissent was more reflective of a progressive labor view, adjusted to incorporate some environmental perspective, than a well-developed environmental critique in its own right. However, the linkage of concerns was well established, particularly in the case of agricultural and military policy. Regarding the former Grossman asked, "Can we suggest models of production and ownership which result in nutritious food, more jobs, less energy and chemical use, [and] less ecological destruction?"[11]

The value of the debate was largely in bringing the de facto American industrial policy out into the open and in highlighting the question of who was doing the picking and choosing and who was benefiting. Although much of the rhetoric was generic, both de facto and prospective industrial policy naturally devolved to sectoral policy, which was itself illuminated by the debate. This illumination was extremely important; however, in the translation to an overt federal policy, industrial (or sectoral) policy could be seen to present extreme risks to the economy, the environment, and the democratic process itself. It tended to represent a centrist approach to real economic and social problems that, in circumventing both democratic and market processes, alienated both ends of the political spectrum, but, to the extent that particular sectors seemed likely to benefit from particular proposals, it also included various supporters from both the left (organized labor) and the right.

Several other elements of the debate are also important in considering different ways to make the translation from sectoral analysis to policy and action. One is the difference between constitutional government authority and, essentially, the bully pulpit. A coherent policy perspective, industrial or otherwise, need not be written into law or even endorsed by the government to have an effect. Indeed, advocates for softer forms of industrial policy were largely arguing for a process of consensus building. This process offered a vitally important opportunity for a more focused public debate around the history and future of particular sectors and associated government policies. The question was and remains, much as Grossman asked in the case of agriculture, Does the environmental community have coherent alternative sectoral perspectives to put on the table? This harks back to the question of strategic vision explored at the end of chapter 3.

At present, there is only one coherent perspective on where the domestic paper industry should be headed in the future, and that is the view held by the industry itself, which has been both passively and actively endorsed by the federal and state governments. The paper industry's vision of the future is one of:

· continued growth in woodpulp production and per capita paper consumption
· accelerated growth in net exports of all pulp and paper commodities
· continued mining of domestic and foreign virgin forests where available, transitioning to intensive plantation forestry as they are exhausted
· incremental growth in wastepaper utilization, possibly augmented by limited growth in nonwood fiber utilization
· incremental improvement in established chemical process and pollution control technology
· continued concentration of production around extremely large-scale kraft woodpulp mills although possibly with continuing peripheral growth in wastepaper-based minimills, and
· continued job loss.

To the extent that this vision is responsive to environmental concerns, it reflects a process of "least steps taken down the path of least resistance" and promises no fundamental change. From a government industrial policy perspective, it is a merger of the old industrial policies reflected in resource management policies, with a technocentric regulatory policy and, lately, with a sectoral technology policy embellishment as represented in Agenda 2020. The latter overtly reflects the elitism and subversion of democratic process that has most concerned progressive voices in the industrial/technology policy debates, even as they acknowledge a legitimate role for government support in areas of basic science and technology development.[12]

Yet, the current sum of objectives held by the extended environmental community around paper recycling, paper procurement, forest use, nonwoods, pulp mill pollution and so on, offers something of a bundle of contradictions rather than a coherent perspective. The vantage gained by a more comprehensive environmental analysis of the sector, however, suggests clear paths for building consensus on the resolution of these apparent contradictions, as well as a clear linkage to issues of social justice and community life. There is, surely, no single trajectory to be endorsed or taken, but common guidelines can be devised to constrain the available options to those that do not cancel out the real prospects for an environmentally grounded transformation of the industry. Some of those guidelines are suggested at the end of this chapter. A longer-term perspective additionally presents the opportunity to reduce the interregional and urban/rural polarization that is often characteristic of different environmental views on the industry. The (subnational) regional perspective has in fact been an important dimension of the industrial/technology policy debates, as explored further below.

Regional Planning

As with the preceding discussion, the intention in this section is to provide some overview of the practical relevance of certain approaches in supporting the opportunity for effective public influence on the evolution of the paper industry. There are, of course, an even larger number of perspectives on regionalism and public planning than there are on industrial policy, and we must begin by approaching them separately.

Planning is fundamentally concerned with the translation from theory and comprehensive analysis to more or less general guidelines for action.[13] The discussion here is essentially focused on planning processes that are relatively ad hoc and informal in nature, although as efforts develop, particularly at community or regional levels they can become more formalized. Types of public sector planning processes that have already arisen in relationship to the paper industry range from organizing efforts focused on particular mills, to initiatives focused on drawing paper industry development into a metropolitan area, to efforts focused on stabilizing existing capacity in in low income urban/suburban areas. Various regional efforts feature interaction between government agencies, university research centers, and nongovernmental public interest organizations.[14] Most often such efforts arise naturally in response to a problem: high unemployment in depressed urban areas, a surfeit of wastepaper, severe local pulp mill pollution, restrictions on the wood supply, and so on. What is suggested here is the potential for such processes when they are further informed by some understanding of the larger, long-range issues attached to the industry and by an awareness of the degree to which local decision making can consciously or unconsciously support or undercut the prospects for both long-range transformation of the industry and maximum local opportunity.

Planning, however, requires an analysis as a platform. The analysis here has been largely focused on outlining, from an environmental perspective, the characteristics of the national industry and its interaction with the global economy via the flow of wood, pulp, paper, and wastepaper across national borders. In the case of a mature basic industry like the paper industry, which is fundamentally national and international in scope and structure, this national perspective is an essential starting point. However, the real potential for change lies in the regional analyses that may follow.

On one hand, different pieces of the picture—fiber and other resources, producers, consumers—obviously exist in a drastically different mixture at different regional levels. Vast regions of the country (the entire Southwest and California for example) are almost wholly dependent on imports of pulp and paper, and on export markets for wastepaper, whereas other regions have functioned as virgin fiber and commodity supply regions. Even from a top-down national view, some regional perspective is necessary to understand the basic equations of the system. Theoretically, in this view, nothing is particularly disturbing about the extreme variations that exist at different subnational levels, as long as new supply regions can be found to replace exhausted supply regions.

In practice, however, national debates—particularly in a nation as large and complex as the United States—are fundamentally concerned with reconciling or choosing between different regional perspectives. In analyzing the industrial policy debates, for example, Robert Kaus wrote of the appeals for "regional equity" intertwined with the *preservationist* approach (backing the losers) to the issue. He, like most, tended to favor the logic of the *accelerationists* (those focused on high-tech sunrise industries) due to the likelihood that advanced technologies would produce a larger national payoff as new locomotives of economic growth. He argued, "Many of the 'social costs' of economic decline in the Northeast, after all, are balanced by the 'social benefits' of economic growth elsewhere. America is a nation; the Northeast isn't." (Actually, he was arguing rhetorically and favored neither approach, suggesting that the democratic process itself was most in need of repair.)[15]

This, however, is the top-down perspective on regions and industries. The bottom-up perspective, the one that originates from within a particular region, is the one of interest to this discussion. There is no alternative here but to do injustice to a literature on regionalism that reaches far back into history and has rarely been less than vital and impassioned; characteristics that suggest the animating power of the framework.[16] It is also clear that regionalism is inextricably intertwined with environment—with the unique natural endowments and contours of a particular place. This connection is proactively embraced in the evolving literature of bioregionalism.

Bioregionalism, like much of the more searching commentary on sustainability, is often characterized more by elaboration and justification of the theory and the vision than by strategies. The abstraction and futurism required to think about either a bioregionally organized society and/or an ecologically sustainable economy often translate poorly to opportunities and barriers that characterize the present. As Mark Dowie states more bluntly, "Although bioregionalism makes considerable sense in theory it seems hopelessly romantic at present."[17] One might say the same thing about sustainability. For purposes of this discussion, the general concepts of region and regionalism are more than sufficient; however, the basic approach here, based on environmentally grounded sectoral analysis, may provide opportunities to help further the goals of bioregionalism prior to the emergence of a "biological politics."[18] As argued above, a comprehensive sectoral perspective represents a way to both honor and contain the complexity of industrial systems and the economy, thus making meaningful action more possible. As such, it can both stimulate and be stimulated by the sharper focus offered within the regional framework and the further richness of local detail that emerges.

An interesting effort based on the synthesis of an integrated environmental perspective on the paper industry and a regional development perspective has been undertaken by the nonprofit Bioregional Development Group in the south of England (see chapter 2). Discernable in the perspective of this group, however, is not only the concept of *building up* sustainable regional industrial sectors, but equally the concept of *delinking from* the larger economy. (One is reminded of alternative energy practitioners who speak of going "off the grid.") Stated another way, it is not only about import substitution, but equally about a disinclination toward exports (one might say export substitution).

This perspective, of course, is anathema in the present focus on international competitiveness and export boosting, although both the economic and environmental logic is relatively self-evident. The riskiness of a local (or regional or national) economy heavily dependent on export markets, particularly distant ones, is not difficult to understand. Political and economic instability in distant regions, sharp changes in fuel costs (for transportation), and fluctuations in national currencies can all bring swift and dramatic change. When supply lines are short, these factors are either easier to predict or not relevant.

The cost factors associated with transport are, of course, considerable and are particularly germane to the economics of new production based on local materials oriented toward regional markets. James McAdoo suggests that transport costs currently comprise, on average, 10–25 percent of the delivered cost of pulp and paper products. The International Institute for Environment and Development (IIED) has found that transport costs can account for as much as a third of the price of incoming fiber, and that outgoing product transport costs can range from 7–19 percent of the delivered cost of the product. Delivery costs alone may thus be equal to or even greater than cost components such as labor, materials, or chemicals. This will likely be one of the important leverage points for smaller, regionally oriented production formats in general and it may become even more significant with increased environmental pressure for measures like carbon taxes.[19]

The environmental advantage of shorter supply lines has not yet been well articulated. Perhaps more to the point, the environmental disadvantage of long supply lines has been substantially underplayed, and the external costs (air pollution, global warming, etc.) remain uncounted. The following points are salient: (1) transport is the fastest growing source of air pollution in both the developed and developing worlds; for air pollutants such as carbon monoxide and carbon dioxide it is already the single largest source; (2) the transport intensity of development is growing, with a progressive shift toward road and air freight (air transport is the most energy intensive and polluting way to move freight, followed by road, rail, and waterborne transport); and (3) transport is the largest single source of noise pollution, and road transport alone represents a significant mode of land use (occupying up to one-third of urban land space in the United States).[20] Needless to say, these factors are particularly relevant in the case of basic materials products such as pulp and paper whose value is commonly measured per ton of product (unlike, say, high-value-added microelectronic components). McAdoo estimates pulp and paper freight at slightly less than 10 percent of world trade. It should also be emphasized that the negative environmental impacts of transport extend equally to wastepaper. There are both powerful economic and environmental bases from which to reject export boosterism or export-driven economic development, particularly when basic materials are involved.

The merits of import substitution—cultivating regional industries to supply regional markets now dependent on imports—are obvious even within the context of dominant economic premises and are at the heart of many regional industrial planning perspectives due to their powerful multiplier effects.[21] The bioregional or sustainability perspectives add the additional, nontrivial qualification that whatever is to be produced must be done in a way that is ecologically sustainable within the region.

To simplify some of these perspectives, the following guidelines on trade can be considered consistent with the prospects for developing a sustainable paper industry and with the opportunities available within a regional planning framework:

- New paper industry development, including the expansion of existing mills and the reopening of idle mills, should not have a long-term orientation toward export markets or a heavy dependence on imported materials, but rather should be oriented toward serving regional markets using regionally available materials. The same general guideline applies to efforts to develop or redevelop secondary and raw materials markets: regional market development should be preferred and heavy export dependency discouraged.
- Existing export-dependent production capacity (including long-distance interstate and intracontinental export dependency) is both economically and environmentally undesirable.

Because our present context is one of subnational regions of the United States and of the paper industry, the latter point speaks to a significant proportion of the established base of production, particularly the largest pulp mill complexes. For practical purposes, assuming a fiber supply to these mills remains available, there is little opportunity on short-term horizons. However, a longer-range perspective may suggest both local opportunity and motivation to discourage or rechannel major recapitalization of long-distance export-dependent production capacity. (Rechanneling may refer, for example, to the establishment of local funding sources for economic diversification programs, as well as to the cultivation of regional markets).[22] Neither approach suggests that total regional self-containment is desirable. Rather, they point to the need to identify the existing degree of fiber, pulp, paper, and wastepaper export or import dependency—which will often be extreme even when viewed within large regional boundaries—and to acknowledge this as an important element of the prospects for sustainable paper industry development. When one considers the various fiber sources (wood, nonwoods, wastepaper), there is likely no major region of the country that lacks fiber resources (or the potential to develop them) sufficient to supply a large proportion of its pulp and paper requirements.

With respect to existing or prospective mill capacity, the general approaches available in the public sector are *opposition, endorsement,* and *cultivation.* The last, of course, is the most challenging because it evokes the most need for analysis and planning. The techniques and tools of opposition and endorsement are clear, even if they have often been missing a solid strategic orientation. The techniques and tools of cultivation are essentially unlimited and run much deeper. They suggest endless collaborative opportunities in the context of a sectoral focus, precisely because such a focus naturally embraces a wide range of human concerns.

One area of significant regional opportunity lies in the integration of production-focused interests with a subject that has been largely absent from this analysis: consumption, particularly as it pertains to the role of industrial, institutional, and commercial consumers of paper. Although it has been little addressed here, the consumption side of the equation has been extensively targeted in the larger environmental debates over paper and the paper industry. Consumer opportunities are generally recognized in terms of: (1) source reduction or, as elsewhere characterized, "paper use efficiency,"[23] and (2) the incorporation of environmental criteria into purchasing decisions. The latter has been primarily pursued within the rarified abstractions of life-cycle analysis. A regional perspective, however, suggests additional opportunities reflected in a familiar maxim: Buy Local.

Unaddressed, thus far, is the question of what is a region. The question is best answered by those who might wish to take the initiative to develop a regional perspective on a sustainable paper industry. Also, multiple regions of different scales may be simultaneously applicable. It was evident, for example, in the case of the proposed West Sacramento mill examined in chapter 5, that multiple regional contexts immediately suggested themselves. The opposition was necessarily targeted at the city level, although it drew heavily from wider Sacramento Valley opposition and ultimately from opposition located as far away as Los Angeles. The most troublesome local environmental impacts (i.e., excluding the long-distance water supply issues) cut across several cities, two counties, two air pollution control districts, and one water quality district. The Sacramento Valley was really the major region involved; however, the complexity of jurisdictions and the clear relationship between the project and the state's solid waste policies tended to point to a state framework. In the more proactive modes that emerged, as some players began to envision possibilities for desirable paper industry development, an immediate linkage was made with rural (agricultural and forest) regions north of the city, although there was no real sense of identity with other major metropolitan areas in the state, with the Sierra counties to the east, or with the agricultural counties of the southern Central Valley.

Interestingly, in both the case of the recycling mill and present efforts around rice pulping prospects, there has been a very strong sense of relationship to British Columbia and Alberta. Canadian players have been central in both cases, and the issues of logging in western Canada have been highly visible. In the more recent case, the importance of developing support from large corporate players (potential customers) with demonstrated commitments to environmental issues has begun to bring Los Angeles more into the equation, but the more organized environmental support has primarily emanated from the San Francisco Bay area. The state, as a whole, clearly emerges as a logical framework for analysis and planning, although it must also be understood within the western North America and Pacific Rim contexts, as well as with respect to the dynamics unique to different regions within the state.

Hawaii, on the other hand, presents no such complexities. The state provides the framework, and the analysis has essentially been obvious for several decades. Other regions of the country—the Pacific Northwest, the Northern Rocky Mountain states, the Lake states, the current "woodbasket" states of the South, the urban regions of the Northeast, and so on—all have wildly different dynamics and relationships to the present system of paper production and consumption. Environmental and sustainable economic development voices in those areas, to the extent that they may wish to participate more profoundly in the project of sustainable paper industry development, will have to determine for themselves if a regional planning perspective is of interest or if the tasks are adequately defined within the context of specific mills and communities.

Overall, the strengthening of regional environmental perspectives and of associated organizational and analytical infrastructure shows significant promise. It suggests a period of turning back, or maybe coming home, after decades of environmental policy and strategy overwhelming focused on Washington, D.C. These efforts have produced many successes, but have failed to come close to fulfilling their promise: the magnitude of the solution has not been commensurate with the magnitude of the problem. An increasing number of analysts of environmental policy and environmentalism (not to mention national electoral politics) have recognized that a shift is underway. The intense localism of the antitoxics movement of the 1980s has been one sign; the strengthening critique of the monopoly held by mainstream, nationally focused environmental groups is another. The bridge between the community and the nation, however, is the region. It represents an alternative environmental focal point that may be particularly relevant to efforts organized around basic industries and basic materials flows. It does not suggest that federal policy is unimportant, but that local and regional policy is equally important and so deserves more serious attention.

Information-Oriented Regulatory and Nonregulatory Tools

In the absence of fundamental change in economic principles and indicators, the need for further and better environmental regulatory support—a central mechanism for dealing with failures that arise out of the segregation of environment and economy—is as compelling as ever. Yet, in the United States, the near-term prospects for enhanced support are presently dim, at least insofar as the establishment and enforcement of more precautionary and/or wholistic standards are concerned. However, standards setting is only one of various possible regulatory modes. It is useful to think in terms of the differentiation that Stephen Breyer has made between *classical* and *alternative* modes of regulation. The classical modes create, in his words, "a detailed web of affirmative legal obligations": they essentially tell the subject of regulation what to do. The alternative modes are more oriented toward altering the preconditions of competition than to specific remedies for defects in the marketplace. Disclosure, or information provision, is one of the most important alternative regulatory modes (pollution taxes are another). Those who appreciate the power of market mechanisms and the legitimate role of good information in efficient markets, as well as those who protest the inflexibility associated with classical "command and control" approaches, should find this alternative comparatively attractive.[24]

One of the most important lessons driven home in the antitoxics efforts of the 1980s was, quite simply, the power of information. The chemical disaster in Bhopal, of course, stood in a class of its own in terms of conveying information. Like previous chemical catastrophes, it had an unmistakable impact on ensuing policy directions. Arising out of worker right-to-know efforts in the 1970s, the community right-to-know movement of the 1980s was strengthened by the message of Bhopal and led to the provisions in the Emergency Planning and Community Right-to-Know Act (EPCRA) of 1986 that established the Toxics Release Inventory and related disclosure mechanisms.[25]

It is by now well understood that the TRI has been one of the most powerful forces in the development and growth of pollution prevention policy and practice. Among other things, it produced a range of outcomes that had not been widely predicted in the debates leading to its creation.

Perhaps most important was that, although framed in terms of community right-to-know, it was equally of value to industry itself and to environmental regulators, who have extensively incorporated TRI data into their activities. Although the first releases of TRI data undeniably represented a public relations nightmare for many manufactures, they also represented the first time that many had attempted to account systematically for a wide range of emissions, thus creating the first relatively comprehensive portrait of their own chemical-specific pollution profiles.

Another outcome not widely predicted was the self-enforcing quality of the reporting requirement. Because of the accessibility of TRI reports (they are easily obtained from Federal Depository Libraries, through the EPA's EPCRA hotline, and now on the Internet), they were widely used within the context of community organizing efforts focused around particular facilities. They were also used in hundreds (by now thousands) of analyses from perspectives as diverse as land-use planning and occupational health. As specific questions and discrepancies emerged—such as the inability to find reports for a particular facility, dramatic changes in consecutive years of a facility's reports, or sharp differences in the reports from regional facilities of like type and scale (e.g. oil refineries, or kraft pulp mills)—they functioned as a monitor on accuracy and completeness. The power of the disclosure approach has been reflected in other laws stemming from the right-to-know movement such as California's Proposition 65,[26] as well as in environmental impact reporting requirements established under the National Environmental Policy Act and various state laws.

In terms of the paper industry, the significance of recent and pending expansions of TRI reporting requirements is unmistakable.[27] Some of the proposed expansions, such as the materials accounting approach that requires not just chemical releases but also chemical use to be reported, have already been demonstrated in states such as Oregon and New Jersey. In Oregon, reported annual chemical use in some pulp mills (in 1991) was greater than emissions of reportable chemicals by factors that range as high as 800.[28] Individual mills use tens of million of pounds (each) of such ubiquitous chemicals as chlorine, chlorine dioxide, and sulfuric acid, and tens and hundreds of thousands of pounds (each) of various other heavy metals and solvents. Among other things, the figures significantly highlight the chains of chemical transport that underlie production. The 1995 reporting year for the TRI will reflect the substantial expansion (by more than 280 chemicals, or nearly double) of reportable chemicals. It is expected to provide further illumination not only of the total magnitude of the national toxics problem, but also of the role of the paper industry in it. Other proposals have suggested the inclusion of criteria air pollutants. Although the aggregate contribution of pulp mills to conventional air pollution is relatively well understood based on monitoring and assessment programs established under the Clean Air Act, the wide availability of facility-specific data would be uniquely valuable.

Disclosure-oriented regulatory approaches, whether focused on production sites or products, have often been driven by issues of human health risk. Because it included the possibility of listing chemicals due solely to adverse environmental effects, the TRI further opened the door to wider considerations.

At the other end of a spectrum of information-oriented tools are a host of softer regulatory and nonregulatory approaches targeted toward providing various types of environmental information on products. These approaches include the full range of practices loosely contained under the umbrella of *green labeling*.[29] The green label approach is generally oriented toward environmentally inclined consumers (as opposed to communities or workers) and often done in conjunction with independent certifiers (who may, as in Europe, be regulated or certified by some level of government). Another aspect of product labeling involves the regulation of advertising claims, although in the case of paper products this has thus far been unregulated in the United States. The food labeling standards promulgated in recent years by the U.S. Food and Drug Administration (FDA) reflect both the regulation of advertising claims (e.g., "low fat") as well as a proactive effort to provide consumers with the additional information necessary to make choices informed by multiple health considerations. The latter represents an example of multiple issue disclosure discussed further below.

As with almost any other product, consumers of paper sit at the summit of very complex networks of activities. They have been buffeted by gales of well-intentioned, often conflicting or irreconcilable advice:

buy recycled; buy tree-free; buy TCF, ECF, CCF (clear-cut free); buy lower-brightness paper; buy lower-basis-weight paper (but not if it jams in the copiers, and/or buy new copiers); buy mechanical pulp-content office paper (uses less trees); don't buy mechanical pulp-content office paper (complicates recycling programs); buy local; buy American; buy union; buy less; support innovative mills; support progressive distributors; boycott environmental villains; negotiate with your suppliers; negotiate with their suppliers, etc.

Some distinction is possible between green labeling approaches in terms of (1) those oriented toward providing accurate, basic information on selected aspects of a product's content and other characteristics (and thus, to some extent on its production history), and (2) those oriented toward providing summary conclusions such as a report card or seal of approval based on the assessment of multiple production factors. The advantage of the first is that consumers, particularly those deliberately operating in relation to their regional contexts, may legitimately make different judgments about the importance of different factors. In the absence of products that satisfy multiple environmental objectives, one may have good reason to place more or less emphasis on reducing chlorine use than on increasing nonwood fiber use. The advantage of the second (summary) approach is that more comprehensive and more technical factors can be included. There is, for example, no simple way to summarize the energy or chemical use or emissions factors associated with a specific paper product, nor to do so without incorporating some level of judgment in the weight assigned to different factors. Summary approaches can increase the breadth and depth of considerations, although the consumer is then reliant on the expertise and perspective of the organization that renders judgement. Although this overview is primarily intended to highlight both the complexity of product labeling issues and the importance of further progress in this area, it does suggest three major avenues that can be pursued with respect to more comprehensive product information.

It seems increasingly advisable that a basic multiple issue disclosure mechanism, perhaps analogous to the approach of the FDA's food labels, should be explored for pulp and paper commodities to provide near-term assistance to consumers for at least some product categories. TCF and ECF are widely understood, valid criteria. Total material content by fiber type (e.g., total wastepaper, postconsumer wastepaper, wood, nonwood fiber) and "other" (e.g., fillers) is also of immediate relevance. Basis weights are increasingly well understood as yet another indicator of environmental opportunity. A systematic presentation of these data elements alone would provide an enormous improvement over the level of information presently available to all but the most sophisticated larger consumers.

Various third-party eco-labeling schemes (including those incorporating certification of forest management) should and will continue to develop independently, and compete for recognition and endorsements by governments and other interest groups. They are natural partners to, but not substitutes for, a basic disclosure mechanism.

Consumers, particularly government institutions and larger consumers will continue to have significant opportunity to pursue more negotiated or focused approaches with distributors and manufacturers.

The Paper Task Force, convened by the Environmental Defense Fund, has articulated elaborate guidelines for direct negotiations between large, highly sophisticated purchasers and pulp and paper companies. In relation to sustainable forest management (SFM) concerns, for example, they suggest that large purchasers obtain sufficient product history information from their suppliers to demonstrate a preference for paper made from wood produced in accordance with the following guidelines: AFPA sustainable forestry guidelines; "all applicable" mandatory and voluntary "Best Management Practices;" "adaptive management" (keeping abreast of and responding to new environmental information about forest management); full external stakeholder participation; conservation of biodiversity; and conservation of ecologically important or declining communities. In relation to pollution concerns, they suggest that purchasers obtain directly from mills a report on their state permit requirements, emissions data, and statistical process variability with respect to BOD, color, fresh water use, total energy use, purchased energy use, SO_2, NO_x, total reduced sulfur compounds, bleach plant effluent flow, AOX, COD, and dioxin. All these can be compared to the Paper Task Force's assessment of standard industry performance along these parameters.[30]

What this suggests is a rather daunting new in-house, multiple-issue investigation and synthesis program, based on detailed, mill- and wood source-specific technical and organizational information. The underlying issue they are addressing (essentially by targeting the ability of large players to eliminate intermediate "noise" in the market), however, is one recognized in the sustainable forest products matrix devised by the MacArthur Foundation: the present inability of the market to transmit good information back and forth over the long chains of relationships that stretch from resources to end consumers. At the resource management level, for example, the MacArthur team identified a lack of biological inventories, a lack of regional perspective on social/ecological interaction, a lack of communication between forest researchers and forest managers, and a lack of knowledge about specialized market opportunities for underutilized wood species. At the primary processor (e.g., pulp mill) level they observed a lack of knowledge of available alternative wood species and lack of awareness of the nuances of end-user needs. Secondary processors (e.g., converters) have few if any information links to harvesters and resource managers as well as a lack of market intelligence about SFM demand from end users. Wholesalers (commodity brokers, freight handlers, etc.) distort price information (and thus environmental signals) from resource and producer levels, and they don't know anything about SFM. Retailers and end users say one thing (save trees) and do another (buy on cost criteria), and they don't really know anything about the social, economic, and ecological forces that underlie the rest of the chain.[31]

One is reminded, once again, of both the power of information and the power—the profound multiplier effects—of substantive, comprehensive information proactively made visible and publicly accessible. The strategies suggested by the MacArthur team are in fact heavily oriented toward information mechanisms and particularly highlight the importance of information exchange within regions. The strategies suggested by the Paper Task Force are also heavily dependent on access to good information, although in their present form they say little on the subject of public sources of such information or on the significance of the regional context. (It should also be observed that neither team offered guidance on the subject of nonwood alternatives). Overall, however, the project of effecting substantive redirection of the present trajectory in the paper industry, or in the forest products industries as a whole, will need to aggressively pursue both regulatory and nonregulatory mechanisms for public disclosure and information diffusion.

Guidelines for Change

The core proposition in this chapter is that comprehensive, informed, alternative visions are both possible and necessary to influence the prospects for a sustainable system of paper production and consumption. The following, which may be contrasted with the dominant industry view of the future outlined above, suggests more socially and environmentally appropriate guidelines for progress:

- diversification of the resource base around regionally sustainable fiber sources
- diversification of geographic distribution
- smaller scales of production
- clean production processes optimized for alternative fiber resources and smaller mill formats
- declining per capita consumption, and
- declining long-distance trade in fiber, pulp, paper, and wastepaper.

Of these, the driving force lies in the diversification of the resource base. As discussed in preceding chapters, the increasing scale of production and regional concentration proceeded hand in glove with the homogenization of the fiber supply, which occurred over the course of the past half century. The highly integrated and capital-intensive formats that resulted have supported the resistance to alternative materials use and to more than incremental change in dominant technologies. And, of course, large-scale mass production went hand in glove with mass consumption. In our preoccupation with the power of demand to shape production, we have tended to downplay the degree to which the needs of production (not to mention advertising) can shape consumption. Yet, few who follow the post-World War II history of mass production will miss the emergence of what Jennifer Seymour Whitaker calls "a peculiar twist on the old idea," in which consumer demand now needed to be pumped up so it could absorb continuously expanding supply. As elsewhere stated, "The way to end glut was to produce gluttons."[32]

The "national bargain" with producers, as described by Robert Reich, was that government would minimize its intervention in management prerogatives and producers would continue to provide more jobs.[33] Even on its own terms, the bargain has long since come unglued with the phenomenon of jobless growth: it has been more than a quarter century since the primary processors in the paper industry created a new job, although production has increased by more than half. In environmental terms, the national bargain was predicated on bank robbery.

The question before us is not whether the paper industry, and the larger system of resource and product use it has both shaped and been shaped by, must change. It must, and a process of marginal change has already begun in response to environmental concerns. The real question before us is more difficult and involves fundamental social choices about the dimensions and directions of change. As communities concerned not only with stopping the further escalation of what is clearly harmful, but also with helping to define new directions consistent with long-term goals of social and ecological well-being, we must become advocates for far-reaching industrial alternatives.

In the case of the paper industry, we are speaking of an industrial system that accounts for nearly half of one of the most basic materials chains in the economy. It is a chain characterized by levels of biological depletion and degradation, industrial pollution, and solid waste creation that could hardly have been conceived one long lifetime ago. As Daly and Cobb have suggested, we have little need to resort to wild rhetoric because the "wild facts" of environmental trauma themselves constitute a massive assault on accepted dogma.[34]

Yet, the confluence of multiple forces—the recycling movement, the wood-use reduction movement, various other indications of a larger movement toward a "new materialism," the emergence of the pollution prevention paradigm and the strengthening of a precautionary principle, the waxing of a new regionalism, the increasingly apparent chasm between the conventional indicators of progress and the lived experience of diminishing opportunity, the increasingly apparent linkages between diminishing opportunity and the condition of the environment, and perhaps even the dawning of a new millennium—suggest a moment of heightened opportunity: a chance to lay new foundations with the strength to support the goals of sustainability. It will require a willingness to find new ways to organize and measure our efforts in a range of different areas and an unwillingness to settle for a future that is only a minor variant of the status quo.

Notes

Introduction

1. American Paper Institute, cited in William Kovacs, "The Coming Era of Conservation and Industrial Utilization of Recyclable Materials," *Ecology Law Quarterly* 15 (1988), 561, note 163.

2. *Los Angeles Times* (2 February 1989); *Waste Age* (September 1989); and *Wall Street Journal* (23 December 1988), letters to the editor.

3. U.S. Environmental Protection Agency, *Toxics in the Community* (Washington, D.C.: U.S. Government Printing Office, 1989).

4. Garrett de Bell, "Recycling," in Garrett de Bell, ed., *The Environmental Handbook* (New York: Ballantine Books, 1970), 214.

5. Frances Irwin, "Introduction to Integrated Pollution Control," in Nigel Haigh and Frances Irwin, eds., *Integrated Pollution Control in Europe and North America* (Baltimore: The Conservation Foundation, 1990), 10.

6. World Commission on Environment and Development, *Our Common Future* (New York: Oxford University Press, 1987).

7. Gareth Porter and Janet Welsh Brown, *Global Environmental Politics* (Boulder, Colo.: Westview Press, 1991), 30.

Chapter 1

1. Quoted in "U.S. Paper Industry Prepared for Slower Economic Pace, Says Red Cavaney," *Pulp and Paper* (May 1990), 138.

2. On the Rittenhouse mill see James Green, *The Rittenhouse Mill and the Beginnings of Papermaking in American* (Philadelphia: The Library Company of Philadelphia and Friends of Historic Rittenhouse, 1990).

3. Jeffrey S. Arpan et al., *The United States Pulp and Paper Industry: Global Challenges and Strategies* (Columbia, S.C.: University of South Carolina Press, 1986), 3–2.

4. On the history of the domestic industry see, for example, Lyman Weeks, *A History of Paper Manufacturing in the United States, 1619–1916* (New York: The Lockwood Trade Journal Co., 1916), and Green.

5. Historical fiber use figure is from Midwest Research Institute, *Paper Recycling: The Art of the Possible 1970–1985* (Kansas City, Mo.: American Paper Institute, 1973), 3 (a report prepared for the Solid Waste Council of the Paper Industry).

6. Dates of incorporation from Paul Ellefson and Robert Stone, *U.S. Wood-Based Industry: Industrial Organization and Performance* (New York: Praeger, 1984), 18–21. Sales ranks from *Lockwood-Post's Directory of the Pulp, Paper, and Allied Trades* (San Francisco: Miller Freeman, 1995).

7. *Lockwood-Post's Directory* (1995).

8. U.N. Food and Agriculture Organization (FAO), *Pulp, Paper, and Paperboard Capacity Survey, 1993–1998* (Rome: FAO, 1994).

9. The first North American recycled market pulp mill was the Newstech mill in Coquitlam, British Columbia. See Jackie Cox, "Western Canada's First Recycled Market Pulp Mill Is Coming Online," *American Papermaker* (August 1991), 22.

10. The U.S. *Census of Manufactures* does differentiate between integrated and unintegrated paper and paperboard mills in some parts of its Industry Series (by using six-digit SIC codes), although most data collected and published by SIC codes does not.

11. Observations on mill capacity trends are drawn generally from the trade literature and specifically from *Lockwood-Post's Directory* (various years). A factor of 350–360 operating days per year is typically used to convert between tons per year (tpy) and tons per day (tpd), although lower values may be used for older mills. Tons refers to short tons. In the case of market pulp mills the figure refers to dried pulp and in the case of integrated mills to finished paper or paperboard.

12. On the 1991 expansion of the Aracruz Cellulose mill in Espirito Santo see Marcello Pilar, "World's First One Million-tpy Market Pulp Mill Begins Operations," *American Papermaker* (March 1992), 48. For a less sanguine view of the facility see *Greenwashing* (Washington D.C.: Greenpeace, 1992), a report prepared in advance of the 1992 U.N. Conference on Environment and Development, and Meredith Persily, *Aracruz Cellulose S.A.: A Brazilian Kraft Pulp Company Confronts Sustainable Development,* senior thesis, Center for the Comparative Study of Development, Brown University, 1994.

Shortly before UNCED, in a May 1992 report entitled *Changing Course,* Aracruz Cellulose had been singled out for recognition of excellence by the Business Council for Sustainable Development (now the World Business Council for Sustainable Development) for its internal environmental policies and efforts in sustainable forestry and economic development. During the conference, however, the Greenpeace ship Warrior II staged a thirty-six-hour blockade at the port of Aracruz in protest of what it called "greenwashing" by the company. The event exposed a history of regional opposition to the mill and a stark picture of both the pollution generated by the mill and the environmental desolation of the vast eucalyptus monoculture plantations from which the mill draws its fiber.

13. The vocabulary used here to describe scales of production is roughly the following: very small (5–150 tpd), small (150–300 tpd), moderate (300–600 tpd), large (600–1,200 tpd), very large (1,200+ tpd).

14. On minimills see Robert Kinstrey, "Low-Tonnage Urban Minimills May Be The Wave Of The Future," *American Papermaker* (April 1992), 22–23, and Ken Patrick, "Sharper Focus on Customer Boosts Industry's Interest in Mini-mills," *Pulp and Paper* (December 1994), 75–79.

15. The real mill count is difficult to determine. *Lockwood-Post's Directory of the Pulp, Paper, and Allied Trades* is an annual catalogue of production facilities in the United States and Canada published by Miller Freeman, who also publish *Pulp and Paper, Pulp and Paper International,* and other leading trade journals. The 1995 Directory reports 599 paper(board) mills (integrated and unintegrated) and 347 pulp mills, about the same as in 1992. The 1992 U.S. *Census of Manufacturers* (preliminary report, industry series for pulp, paper and paperboard mills) reports 44 pulp mills, 291 paper mills, and 206 paperboard mills, up from 39, 282 and 205 in 1987. The more detailed data available from the final census tally for the 1987 reporting year identified, in addition to 39 standalone pulp mills, 46 paper mills and 91 paperboard mills that were integrated with a pulp mill.

The census data thus indicates a count of around 175 pulp mills and 500 paper(board) mills, missing some 175 pulp mills and 100 paper(board) mills counted by Lockwood-Post. Although the definition of a pulp mill may be ambiguous when wastepaper is involved, Lockwood-Post identifies more than 250 *woodpulp* mills (still around 75 more than the 1987 census total for all pulp mills), which are relatively unmistakable. Lockwood-Post counts idle mills, whereas Census data counts only those that have a payroll during the year; yet, the low census numbers are still hard to reconcile. Total employment counted by Lockwood-Post's Directory is also substantially at odds with census data as described later. U.S. Dept. of Commerce, Bureau of the Census, *Census of Manufactures, Industry Series: Pulp Mills, Paper Mills, and Paperboard Mills* (Washington, D.C.: U.S. Dept. of Commerce, Bureau of the Census, 1987, 1992-preliminary).

16. *Lockwood-Post's Directory* (1995).

17. From the matrix/poster entitled "Sustainable Forest Products: Opportunity within Crisis" from the John D. and Catherine T. MacArthur Foundation, Chicago, Illinois, 1995.

18. See Henry Spelter, *Capacity, Production, and Manufacturing of Wood-Based Panels in North America* (Washington, D.C.: U.S. Dept. of Agriculture, Forest Service, 1994).

19. U.S. Dept. of Commerce, Bureau of the Census, *Census of Manufactures, Industry Series* (volumes associated with sectors SIC 24 and SIC 25) (Washington, D.C.: U.S. Dept. of Commerce, 1987, 1992-preliminary), and *Census of Manufacturers, Summary Series* (U.S., industry and geographic areas statistics) (1992-preliminary).

20. Christopher Kelly, "The Southern Pulpcutter and the 'Short Stick': The Mississippi Uniform Pulpwood Scaling and Practices Act," *Clearinghouse Review* 19 (1986), 1153–56.

21. Ellefson and Stone. Ellefson and Stone provide a much more comprehensive portrait of the forest products industry than is presented here. Although the statistics have become dated, the basic organization they describe is still relevant and the book provides a useful guide for understanding the industry, as well as a model for studying its structure.

22. Clayoquot Rainforest Coalition press kits and miscellaneous literature distributed by Rainforest Action Network and Greenpeace (San Francisco, 1994–1995).

23. *Ward's Business Index of U.S. Private and Public Companies* (1992); various corporate annual reports.

24. *Lockwood-Post's Directory* (1995); corporate annual reports; "Weyco Plans Global Timberlands Joint Venture," *Pulp and Paper* (June 1995), 21.

25. Christopher Freeman and Geoffrey Oldham, "Introduction: Beyond the Single Market," in C. Freeman, M. Sharp, and W. Walker, eds., *Technology and the Future of Europe: Global Competition and the Environment in the 90s* (London: Printer Publishers Ltd., 1991), 4–5. Reilly is quoted in Charles Swann, "The New Environmentalism," *American Papermaker* (July 1992), 26.

26. Freeman and Oldham.

27. Project on Industrial Policy and the Environment, *America's Economic Future: Environmentalists Broaden the Industrial Policy Debate* (New York: Natural Resources Defense Council, 1984). The Steering Committee for this project consisted of representatives from Natural Resources Defense Council, the Wilderness Society, Friends of the Earth, Sierra Club, National Audubon Society, Environmental Defense Fund, Public Citizen's Critical Mass Energy Project, Environmental Action, Urban Environmental Conference, and Environmentalists for Full Employment. A vigorous dissent by Richard Grossman of Environmentalists for Full Employment was joined by several other participants as discussed in chapter 6.

28. National Science and Technology Council, *Technology for a Sustainable Future: A Framework for Action,* executive summary (Washington, D.C.: NSTC, 1994).

29. Nicholas Ashford and George Heaton, Jr., "Regulation and Technological Innovation in the Chemical Industry," *Law and Contemporary Problems* 46(3) (summer 1983), 157.

30. Donald A. Hicks (ed.), *Is New Technology Enough?: Making and Remaking U.S. Basic Industries* (Washington, D.C.: American Enterprise Institute for Public Policy Research, 1988), 2–3.

31. Ashford and Heaton.

32. William Abernathy and James Utterback, "Patterns of Industrial Innovation," *Technology Review* (June-July, 1978), 41, cited in Ashford and Heaton, 110–13.

33. Ronald Slinn, "The Paper Industry and Capital," *Tappi Journal* (September 1992), 16. U.S. *Census of Manufacturers* (1987, 1992-preliminary).

34. Production figures are from U.S. Dept. of Commerce, Bureau of Economic Analysis, *Business Statistics, 1963–91* (Washington, D.C.: U.S. Dept. of Commerce, Bureau of Economic Analysis, June 1992), and from *Lockwood-Post's Directory* (annual). Employment figures are from U.S. *Census of Manufacturers* (1987, 1992-preliminary). Lockwood-Post's Directory reports consistently higher employment figures, for example a total employment figure of 687.9 thousand in 1991, against a census figure of 667.2 thousand.

35. Kraft pulp mill capacity calculated from *Lockwood-Post's Directory* (annual).

36. See Peter J. Ince, *Recycling and Long-Range Timber Outlook: Background Research Report, 1993 RPA Assessment Update, USDA Forest Service* (Madison, Wis.: U.S. Department of Agriculture, Forest Service, Forest Products Laboratory, 1994), 66. Ince provides estimated costs of new capacity (per ton of output) for more than ninety different types of pulping processes. Figures (in 1986 dollars) are lower than those derived here from trade literature.

37. See, for example: Alabama River Newsprint, 220,000 tpy deinking plant ($35 million, approximately $283,000/tpd), *Paper* (June 1992), 11; Champion International, 400 tpd 100 percent recycled newsprint pulp plant at Sheldon complex, east of Houston ($85 million, $212,500/tpd), *Chemical Marketing Reporter* (May 1992), 42; Daishowa America, 200 tpd deinking plant (directory grade) in Port Angeles, Washington ($40 million, $200,000/tpd), *American Paper Journal* (June 1992), 11; Abitibi-Price, 55,000 tpy deinking plant (directory grade) in Alma, Quebec ($20 million, about $125,000/tpd), *Pulp and Paper* (June 1995), 21; Ponderosa Fibres with Boise Cascade, 470 tpd deinking plant (printing paper grade) at Boise's Wallula Washington mill ($140 million, $298,000/tpd), *Pulp and Paper* (May 1995), 21; Hagerstown Fiber L.P., 400 tpd deinked market pulp and wastewater treatment plant in Hagerstown, Maryland ($199.8 million, $500,000/tpd), *Pulp and Paper* (December 1994), 17; Intercontinental Recycling Corp., 400 tpd deinked high-grade market pulp mill with oxygen bleach stage in Fitchburg, Mass. ($220 million, $550,000/tpd), *Pulp and Paper* (December 1994), 17; Ponderosa Fibres, 300 tpd recycled newsprint pulp/paper mill proposed for New York ($150 million, $500,000/tpd), *American Paper Journal* (June 1992), 16; Evergreen Pulp and Paper and Fletcher Challenge Canada, 700 tpd recycled newsprint pulp/paper mill in Arizona ($405 million, $579,000/tpd), *Pulp*

and Paper (July 1992), 27; MacMillan Bloedel, 140,000 tpy recycled linerboard mill in Henderson, Kentucky ($100 million, $355,000/tpd) and Repap Enterprises, 400 metric tpd Alcell hardwood pulp mill (Canadian $300 million, around U.S. $930,000/tpd) (at existing mill site, cost is for new digester and recovery system), *Pulp and Paper* (October 1994), 19.

On announced virgin mills in Alberta, see: *Pulp and Paper* (May 1992), 17 (Alberta-Pacific Forest Industries), and *Paper* (May 1992), 12; and *Pulp and Paper* (June 1992), 23 (Grande Alberta Paper).

38. U.S. National Science Foundation, *Research and Development in Industry* (annual), cited in *Statistical Abstract of the United States* (1994), table 975.

39. American Forest and Paper Association, *Agenda 2020: A Technology Vision and Research Agenda for America's Forest, Wood and Paper Industry* (Washington, D.C.: AFPA, 1994), 3, 6.

40. U.S. *Census of Manufactures* (1992-preliminary).

41. Except where otherwise indicated, all U.S. production, consumption, and trade data (in tons) for pulp, paper, paperboard, and wastepaper is from the following sources: *Lockwood-Post's Directory* (annual); Franklin Associates, Ltd., *The Outlook for Paper Recovery to the Year 2000,* executive summary (Washington, D.C.: AFPA, 1993) (report prepared for the American Forest and Paper Association); Franklin Associates, Ltd., *Paper Recycling: The View to 1995,* summary report (New York: API, 1990) (report prepared for the American Paper Institute, New York, 1990); U.S. Dept. of Agriculture, Forest Service, *An Analysis of the Timber Situation in the United States: 1989–2040,* General Technical Report RM-199, Fort Collins, Colorado (Washington, D.C.: U.S. Dept. of Agriculture, 1990). All world and international paper production and trade figures are from U.N. Food and Agriculture Organization (FAO), *Forest Products Yearbook* (Rome: FAO, 1991).

42. On the Russian paper industry see Kathy Abusow, "Fertile Ground for Investment in Russia's Vast Forests," and "Restructuring and Investment Needed to Waken Sleeping Giant," *Pulp and Paper International* (January, February 1995).

43. Kivoaki Iida, "Current Pulp Production in Japan and its Research and Technological Topics," *Tappi Journal* (April 1995), 67–74.

44. International Institute for Environment and Development, *Non Wood Fiber Substudy,* phase I review report (London: IIED, 1995)—a report prepared as part of a major study of "The Sustainable Paper Cycle" for the World Business Council on Sustainable Development.

45. Some European countries with smaller paper industries (such as Spain, the United Kingdom and The Netherlands) rely on wastepaper for 60–70 percent or more of the fiber used to make paper. Wastepaper utilization rates are from Mahendra Doshi, Gary Scott, and John Borchardt, "Semiannual Conference Review: January-June 1995," *Progress in Paper Recycling* (August 1995), 81. U.S. wastepaper utilization is discussed at length in chapter 4.

46. Arpan et al., 3–2.

47. In table 1.2 international consumption and per capita consumption are calculated from production, import and export data in FAO, *Forest Products Yearbook,* with population and GNP figures taken from the International Bank for Reconstruction and Development (World Bank), *World Development Report 1992* (New York: Oxford University Press, 1992). These figures used to compare U.S. consumption to other countries are, however, significantly lower than published figures from other sources (including the American Forest and Paper Association) used in table 1.3, which place 1991 U.S. per capita paper and paperboard consumption closer to 700 lb.

48. Historical and projected consumption and per capita consumption figures by paper grade are calculated from Franklin Associates, Ltd. (1990, 1993), with population data and projections from U.S. Department of Commerce, Bureau of the Census, *Current Population Reports* (Washington, D.C.: U.S. Department of Commerce, Bureau of the Census, 1993).

49. Claudia Thompson, *Recycled Papers: The Essential Guide* (Cambridge, Mass.: MIT Press, 1992), 10.

50. Ince, 103–104.

51. Claes von Ungern-Sternberg, "Industry Must Fight for Realistic Greening of Trade," *Pulp and Paper International* (January 1995), 55.

52. See Ashford and Heaton, 112.

53. Peter Wrist, "Sustainable Development and Its Implications for the Forest Product Industry," *Tappi Journal* (September 1992), 69 (revised text of the Gunnar Nicholson Lecture presented at the 1992 TAPPI Environmental Conference, 13 April 1992, in Richmond Virginia).

Chapter 2

1. See, for example, W. Michael Hanemann, "Economics and the Preservation of Biodiversity," and Bryan Norton "Commodity, Amenity, and Morality: The Limits of Quantification in Valuing Biodiversity," and others in E. O. Wilson, ed., *Biodiversity* (Washington, D.C.: National Academy Press, 1988).

2. For a classic treatment of the issue see Robert Repetto, "Accounting for Environmental Assets," *Scientific American* (June 1992), 96.

3. Catherine Caufield, "The Ancient Forest," *The New Yorker* (14 May 1990), 48.

4. Perry Hagenstein, "Forests," in Neil Sampson and Dwight Hair, eds., *Natural Resources for the 21st Century* (Washington, D.C.: Island Press, 1990), 88.

5. As International Paper advertises, "We meet the need for virgin fiber through a practice called 'sustainable forestry,' which includes planting 50 million SuperTree™ seedlings every year." Advertisement, *New Yorker* (13 November 1995).

6. Table 2.1 is based on U.N. Food and Agriculture Organization (FAO), *Forest Products Yearbook* (Rome: FAO, 1991).

7. Except where otherwise indicated, all U.S. production, consumption, and trade data for roundwood and lumber from the 1920s through the mid-1980s is from the following sources: U.S. Dept. of Agriculture, Forest Service, *An Analysis of the Timber Situation in the United States: 1952–2030,* Forest Resource Report no. 23 (Washington, D.C.: U.S. Dept. of Agriculture, 1982); and U.S. Dept. of Agriculture, Forest Service, *An Analysis of the Timber Situation in the United States: 1989–2040,* General Technical Report RM-199, Fort Collins, Colorado (Washington, D.C.: U.S. Dept. of Agriculture, 1990). 1990s figures are from FAO (1991).

8. See USDA (1982), 77–83, and USDA (1990), 82–84 and elsewhere for discussions of log exports.

9. Kelly Ferguson, "Northwest Logging Limits Drive Forest Product Export Decline," *Pulp and Paper* (October 1994), 63–68.

10. USDA (1982), 292; and USDA (1990), 23–27.

11. FAO (1991). Consumption of roundwood by lumber manufacturers is estimated from 1986 data in USDA (1990), 197.

12. USDA (1982), 493.

13. USDA (1990), 197.

14. Ronald Slinn, "Fiber from Forests—Facts and Figures," *Tappi Journal* (December 1992), 14.

15. *Fibre Market News* (17 August 1990), 1, 14.

16. For a brief treatment of the subject see John Ruston, "Developing Markets for Recycled Materials," Environmental Defense Fund Working Paper no. 43 (New York: Environmental Defense Fund, 1988). Also see the overview provided in U.S. Congress, Office of Technology Assessment, *Facing America's Trash: What Next for Municipal Solid Waste,* OTA-O-424 (Washington, D.C.: U.S. Government Printing Office, 1989), 197–202.

OTA observes that the major areas of subsidy to virgin timber production are: (1) multiperiod expensing of timber growing costs (an exemption from provisions of the Tax Reform Act of 1986 that allows the timber industry to distribute interest for capital costs over the entire production period); (2) a 10 percent tax credit for direct investments in timber stands and a provision allowing the direct costs incurred in reforestation to be amortized over a seven-year period; (3) below-cost government timber sales (the sale of timber from federal lands at below-market—and below administrative costs—value, which has been based on the argument that the purchaser pays to build logging roads that can then be used by everyone; and (4) technical support from the Department of Agriculture. The first two were roughly estimated to cost the federal government $474 million in fiscal year 1989. Randall O'Toole provides a detailed analysis of below-cost timber sales in *Reforming the Forest Service* (Washington D.C.: Island Press, 1988).

The impacts of these forms of subsidy are only poorly understood. The OTA, for example, could not explain why the government cost of the multiperiod expensing provision jumped significantly after special capital gains treatment of income from timber production was eliminated in 1988 (OTA, 199). In general, however, OTA observes, "These programs have been used for decades to stimulate, and sometimes help maintain, these sectors of the economy. As a result, the programs have become embedded in the economic system and are now an integral part of the industrial infrastructure and economics of natural resource development and production" (OTA, 197).

17. Caufield, 46.

18. Robert W. Hagler, "The Global Wood Fiber Equation—A New World Order?" Proceedings, Third International Pulp and Paper Symposium: *What is Determining International Competitiveness in the Global Pulp and Paper Industry?* 13–14 September 1994 (Seattle: University of Washington, Center for International Trade in Forest Products, 1994) (hereafter CINTRAFOR Proceedings), 269.

19. Council on Environmental Quality, *Environmental Quality 1993,* 24th annual report of the Council on Environmental Quality (Washington, D.C.: Office of the President, CEQ, 1993).

20. U.S. Dept. of Commerce, Bureau of the Census, *Census of Manufactures, Industry Series: Logging, Sawmills, and Planing Mills* (Washington, D.C.: U.S. Dept. of Commerce, Bureau of the Census, 1987, 1992-preliminary).

21. USDA (1990), 113–16; and USDA (1982), 496–99 (glossary of terms).

22. USDA (1982), 119.

23. See Ronald Slinn, "The Impact of Industry Restructuring on Fiber Procurement," *Journal of Forestry* (February 1989), 17. Also Hagenstein, 86.

24. USDA (1990), 49–51.

25. Caufield, 67–69; also see Hagenstein, 93–99.

26. USDA (1990), 159–72.

27. USDA (1990), 49–51; and Hagenstein, 88. On California see *L.A. Weekly* (7 September 1990), 22.

28. Slinn (1989), 18.

29. *Lockwood-Post's Directory* (1990), 14; USDA (1990), 69; and Clark Wiseman, "Increased U.S. Wastepaper Recycling: The Effect on Timber Demand," *Resources, Conservation and Recycling* (January 1993), 109.

30. Con Schallau and Alberto Goetzl, "Effects of Constraining Timber Supplies: Repercussions are National and Global," *Journal of Forestry* (July 1992), 23.

31. Scott Norvell, "Southern Comfort for a Timber Giant," *New York Times* (28 March 1993). Norvell quotes Steven Corvitt of the Mississippi Forestry Association. Also see Michael Parrish, "Western Debate Hides Timber's Flight South," *Los Angeles Times* (9 February 1992).

32. Hagler, CINTRAFOR Proceedings, 271.

33. For a thorough recent treatment of the subject of timberland investment see F. Christian Zinkhan, William Sizemore, George Mason, and Thomas Ebner, *Timberland Investments* (Portland, Ore.: Timber Press, 1992).

34. "Woodpeckers and Wetlands in the South," *New York Times* (28 March 1993); and Schallau and Goetzl, 25. Also see Ralph Colberg, "Southern Softwood Forests Cannot Support Projected Demand," *American Papermaker* (July 1992), 34.

35. Jared Wolfe and Melody Mobley, "Nonfederal Public Forests," *American Forests* (November/December, 1989), 31.

36. Schallau and Goetzl, 23, 26.

37. On Honduras and Brazil: *Paper* (May 1992), 20. On Chile and Canada: Zinkhan et al., 16. On New Zealand: *Forest Industries* (August 1992), 5. On lumber production in the former Soviet Union and Venezuela: Michael Parrish, "Timber War Branches Out," *Los Angeles Times* (2 April 1993). On the environmental impacts of timber production in Siberia: Schallau and Goetzl, 25.

38. B. Lipke, "Meeting the Need for Environmental Protection While Satisfying the Global Demand for Wood and Other Raw Materials: A North American and Global Trade Perspective," paper presented at the Conference on Wood Product Demand and the Environment, Vancouver, B.C., 12–14 November 1991; and J. Bowyer, "Responsible Environmentalism: The Ethical Features of Forest Harvest and Wood Use," University of Minnesota, St. Paul—both as cited in Schallau and Goetzl, 25. Also comments by Schallau and Goetzl, 25.

39. F. Keith Hall, (untitled), Proceedings, 1992 TAPPI Environmental Conference (Atlanta: TAPPI, 1992)—text of a speech delivered to the Technical Association of the Australian and New Zealand Pulp and Paper Industry (APPITA) in April 1991. TAPPI is the Technical Association of the Pulp and Paper Industry in the United States.

40. Roger M. Powell, "Opportunities for Value-Added Bio-Based Composites," paper delivered at the Pacific Rim Bio-Based Composites Symposium, Rotorua, New Zealand, 9–13 November 1992.

41. Publications lists from the USDA Forest Service's Forest Products Laboratory (Madison, Wisconsin) provide a good introduction to this literature. In addition, there are an increasing number of annual conferences focused on lignocellulosic composites and on engineered or reconstituted fiber products. For introductory overviews see Brent English, John A. Youngquist, and Andrzej M. Krzysik, "Lignocellosic Composites," in Richard Gilbert, ed., *Cellulosic Polymers: Blends and Composites* (New York: Hansen Publishers, 1994); Theodore L. Laufenberg, "Concepts for Fiber-Based Structural Building Systems," Proceedings, International Panel and Engineered-Wood Technology Exposition, Atlanta, Georgia, 19–20 October 1993; and Powell.

42. See, for example, comments by Powell (1992), 245, 250.

43. *The Need for a National Materials Policy: Hearings before the Subcommittee on Environmental Pollution of the Senate Committee on Public Works,* 93rd cong., 2d sess. (11–13 June, 9–11 and 15–18 July 1974).

44. Gottlieb notes, for example, that the 1952 President's Materials Policy Commission focused on the availability of materials and on possible materials shortages rather than on conservation and efficiency of use. In turn, materials security was largely framed in Cold War terms of national security. Robert Gottlieb, *Forcing the Spring: The Transformation of the American Environmental Movement* (Washington, D.C.: Island Press, 1994), 38, 309.

45. John Young, "The New Materialism: A Matter of Policy," *World Watch* (September/October 1994), 30–37.

46. William Cronon, *Nature's Metropolis: Chicago and the Great West* (New York: W. W. Norton, 1994), xvii.

47. On the shift from pollution control to pollution prevention policy, the debates over terminology and focus, and interdisciplinary pollution prevention studies see Robert Gottlieb, ed., *Reducing Toxics: A New Approach to Policy and Industrial Decisionmaking* (Washington, D.C.: Island Press, 1994).

48. For an introduction to the larger chlorine controversy and calls for sunsetting most industrial applications of chlorine and chlorine compounds see, for example, Bette Hileman, Janice R. Long and Elisabeth M. Kirschner, "Chlorine Industry Running Flat Out Despite Persistent Health Fears," *Chemical and Engineering News* (21 November 1994), 12–26. The subject is further discussed in chapter 3.

49. Atossa Soltani and Penelope Whitney, eds., *Cut Waste Not Trees: How to Use Less Wood, Cut Pollution and Create Jobs* (San Francisco: Rainforest Action Network, 1995).

50. Systems Group on Forests, Monthly Updates nos. 1–3 (July, October, December 1995), Sacramento, Calif., Global Futures. The Fortune 500 Roundtable, also a joint project of RMI and Global Futures, includes corporate members committed to "advanced resource productivity." The Wood Reduction Clearinghouse is based in San Francisco.

51. See *Earth Island Journal* (Winter 1993–94), 19, and a series of articles on various aspects of tree-free paper in 1994 and 1995 volumes. The *Journal* is now being published entirely on kenaf paper.

52. U.N. Food and Agriculture Organization (FAO), *Pulp, Paper, and Paperboard Capacity Survey, 1993–1998* (Rome: FAO, 1994); *Lockwood-Post's Directory* (1995).

53. Table 2.3 is based on Joseph E. Atchison, "Nonwood Fiber Could Play Major Role in Future U.S. Papermaking Furnishes," *Pulp and Paper* (July 1995), table 3, 127. The figures are not tied to a specific year, but refer (presumably based on information available as of 1994–95) to fibers that "could be made available."

54. USDA (1982), 298.

55. Atchison (1995); and Atchison, "Present Status and Future Prospects for Use of Nonwood Plant Fibers for Paper Grade Pulps," presentation, 1994 American Forest and Paper Association Pulp and Fiber Conference, Tucson, Arizona, 14–16 November 1994, as cited in Paper Task Force, "Nonwood Plant Fibers As Alternative Fiber Sources for Making Paper," White Paper 13 (New York: Environmental Defense Fund, July 1996), 9.

56. Joseph Atchison, "Historical Development of Bagasse Newsprint Processes Proposed and Mills Built, Present Status of Bagasse Newsprint Technology, and Possibilities for the Future," *Nonwood Plant Fiber Pulping, Progress Report no. 10* (Atlanta: TAPPI, 1979); Joseph Atchison, "Making the Right Choices for Successful Bagasse Newsprint Production: Part I," *Tappi Journal* (December 1992).

57. Recent efforts have focused on both wastepaper and industrial fiber crops such as crotalaria that could be grown on lands vacated by the now-declining Hawaiian sugar cane industry. They have been spearheaded by the Clean Hawaii Center and the Hawaii Dept. of Business, Economic Development, and Tourism. Personal communication with Mary Hieb, MLH Associates, Maui, Hawaii (October 1995).

58. For a summary of USDA efforts on kenaf see Daniel E. Kugler, *Kenaf Newsprint: Realizing Commercialization of a New Crop after Four Decades of Research and Development* (Washington, D.C.: U.S. Dept. of Agriculture, undated). The original series of technical papers describing the screening of more than five hundred plants, and eventually focusing on pulping and papermaking studies of kenaf are available as reprints from the USDA, along with subsequent technical studies of kenaf by the USDA.

59. Personal communication with Tom Rymsza, KP Products Inc., Albuquerque, New Mexico (December 1995); KP Products Inc., *Response to the Paper Task Force White Paper No. 13* (submitted to the Environmental Defense Fund, 13 January 1996, available from KP Products); personal communication with Colin Felton, U.S. Dept. of Agriculture, Forest Service, Forest Products Laboratory, Madison, Wisconsin (January 1996).

60. U.S. Dept. of Agriculture, Cooperative State Research Service, *Kenaf Pulp and Fiber Commercialization Slow and Costly despite Proven Economic and Environmental Benefits* (Washington, D.C.: U.S. Dept. of Agriculture, August 1993).

61. Personal communication with Tom Rymsza, KP Products Inc., Albuquerque, New Mexico (December 1995).

62. Jason L. Merrill, "The Manufacture of Paper from Hemp Hurds," U.S. Dept. of Agriculture Bulletin no. 404 (16 October 1916), partially reproduced in John T. Birrenbach, *A Report on the Use of Cannabis Hemp as a Source of Raw Materials in the Production of Paper: American Grown Hemp Can Supply Our Paper Needs* (St. Paul, Minn.: Greenleaf Publications, 1993). Birrenbach is the founder of the Institute for Hemp.

63. H. J. Nieschlag, G. H. Nelson, I. A. Wolff, and R. E. Perdue, Jr., "A Search for New Fiber Crops," *Tappi* (March 1960), 199.

64. The Colorado Hemp Production Act was introduced into the Colorado State Senate in 1995 and again in 1996, and would redefine "marijuana" to exclude low-THC industrial hemp. Various materials from Colorado Hemp Initiative Project (Nederland, Colorado, 1994–96). In 1994, the Governor of Kentucky set up a task force to examine prospects for industrial hemp cultivation in the state. Todd Pack, "Kentucky Governor Looks into Creating, Legalizing Marijuana-Free Hemp," *Lexington Herald-Leader* (24 November 1994).

65. Stephen Kinzer, "Germany Lifts Ban on Hemp Growing," *New York Times* (12 November 1995).

66. Ed Ayres, "Making Paper Without Trees," *World Watch* (September/October 1993), 7.

67. Frans Zomers, Richard Gosselink, Jan van Dam, and Bôke Tjeerdsma, "Organosolve Pulping and Test Paper Characterization of Fiber Hemp," *Tappi Journal* (May 1995), 149–155.

68. Al Wong, Romuald Krzywanski, and Chen Chiu, "Agriculture-Based Pulp As Secondary Papermaking Fibre," Proceedings, 1993 TAPPI Pulping Conference, vol. 1, Atlanta, Georgia, 1–3 November 1993; Mark Stumborg, Lawrence Townley-Smith, and Ewen Coxworth, "Export Issues and Alternate Uses for Cereal Crop Residues," presentation, 1994 AIC Symposium on Sustainable Cropping, Regina, Saskatchewan, 13–14 July 1994.

69. The rice straw project is being led by Al Wong, of Arbokem Ltd., and was in an exploratory phase in 1995–96. The rye grass straw project was undertaken by Weyerhaeuser in collaboration the USDA Alternative Agricultural Research and Commercialization Center and the Oregon Dept. of Agriculture. Personal communications with Al Wong, Arbokem Ltd., Vancouver, British Columbia (August 1995, May 1996) (rice straw); W. Nay, "Oregon Ryegrass Straw for Containerboard," paper presented at the Pacific TAPPI meeting, 16 September 1994 (grass straw).

Other projects organized around agricultural residues include the work of the Herty Foundation and Heartland Fibers to commercialize corn stalk pulping, and the Texas Grain Sorghum Project as described in David Pettijohn and David Lorenz, *Agricultural Sources of Fiber for Paper Production: Economic and Environmental Benefits,* draft (Washington, D.C.: Institute for Local Self-Reliance, 1996).

70. Walter Bublitz, "The Pulping Characteristics of Straw Residues from Oregon Seed Grass Production," *Nonwood Plant Fiber Pulping, Progress Report no. 2* (Atlanta: TAPPI, 1971).

71. American Forest and Paper Association, *Agenda 2020: A Technology Vision and Research Agenda for America's Forest, Wood and Paper Industry* (Washington, D.C.: AFPA, 1994), 13.

72. In the United States, roughly 88 million tons of wastepaper was generated in 1992, with about 34 millions tons (38 percent) recovered: an amount sufficient

to yield around 27 million tons of pulp (assuming an 80 percent pulping yield). Atchison (1995) estimated that 290 million tons of agricultural residues were also available. A 38 percent recovery rate would provide an amount sufficient to yield roughly 44 million tons of virgin pulp (assuming a low average yield of 40 percent). The U.S. paper industry consumed well over 40 percent of the virgin wood and wood chips used in the United States and produced about 63 million tons of woodpulp in 1993. The pulp yield from wood ranges from roughly 40 to 95 percent depending on the chemical or mechanical processes employed. Franklin Associates, Ltd., *The Outlook for Paper Recovery to the Year 2000* (Washington, D.C.: AFPA, 1993) (wastepaper generation and recovery); Atchison (1994, 1995) (agricultural residue); and *Lockwood-Post's Directory* (1995) (woodpulp production).

73. Judy Pasternak, "Planting Hope and Crambe," *Los Angeles Times* (9 March 1995).

74. Atchison (1995), 130.

75. Lynn Graebner, "Rice Straw May Again be a Burning Question," *Business Journal* (26 June 1996), 3; Vince Bielski, "Paper Panacea," *Sacramento News and Review* (6 June 1996), 20.

76. Andrew Kaldor, "Kenaf, An Alternative Fiber for the Pulp and Paper Industries in Developing and Developed Countries," *Tappi Journal* (October 1992), 141.

77. See Paper Task Force, "Nonwoods," section IV, for comparisons of tree plantations and annual fiber crops; also Pettijohn and Lorenz.

78. Punya Chaudhuri, "Sowing the Seeds for a New Fiber Supply," *Pulp and Paper International* (March 1995), 68–69.

79. Brochure, Association for the Advancement of Industrial Crops (1993).

80. Al Wong, "Canada's New Agriculture for the Third Millennium," submitted to the Canadian Senate-Commons Committee on Agricultural Policy (October 1994).

81. KP Products (1995).

82. Kazuhiko Sameshima, "Japanese Local Paper Mill Needs a Way to Survive: Kenaf is One of the Hopefuls," Proceedings, 1992 TAPPI Environmental Conference (Atlanta: TAPPI, 1992).

83. Penelope Whitney, "Kentucky Hemp Revival: Interview with Farmer James Burton," in Soltani and Whitney (1995), 37.

84. Sue Riddlestone, "Growing Our Own Clothes: Heritage, Ecology and the Future," *Permaculture Magazine* 7 (1994), Riddlestone and Desai, *Bioregional Fibres: The Potential for a Sustainable Regional Paper and Textile Industry Based on Flax and Hemp* (Kent, U.K.: The Bioregional Development Group, 1994); International Institute for Environment and Development (IIED), *Non Wood Fiber Substudy,* draft (London: IIED, 1995), 4, 36.

85. Zomers et al., 150; Chaudhuri, 68; Wong, Krzywanski, and Chiu (1993) (scale); IIED, 26 (economic radius).

86. An even more troubling example of how a pervasive wood-based industy bias shapes the "independent" analysis of nonwoods is found in the International Institute for Environment and Development's (IIED) recent report on the paper industry and its substudy on nonwoods. Unlike the Paper Task Force, IIED clearly failed to benefit from its review process, or to make a serious attempt to contemplate the perspectives communicated in its much-advertised consultations with environmental NGOs. IIED, *Towards a Sustainable Paper Cycle,* final report review draft (London: IIED, 1996); personal communication with environmental NGO representatives at IIED NGO consultation on a Sustainable Paper Cycle, London, 19 May 1995.

87. KP Products, "Response"; personal communication with Emily Miggins, ReThink Paper, Charlotte Fox, Wallace Global Fund, Tom Rymsza, KP Products, and John Ruston and Richard Dennison, EDF (December 1995); personal communication with Lauren Blum, EDF (July 1996). The final summary report of the Task Force (vol. 1), based on a series of white papers (vol. 2), was released in December 1995: Paper Task Force, *Paper Task Force Recommendations for Purchasing and Using Environmentally Preferable Paper* (New York: Environmental Defense Fund, 1995). It should be noted that the revised final report was printed on a paper containing 45 percent wheat straw, 40 percent recycled pulp, and 15 percent chalk.

88. Kaldor, 145.

89. Jerry Powell, "$12 Million Homes and Cheap Trees," *Resource Recycling* (November 1995), 6 (citing a recent study from the U.S. Government Accounting Office on below-cost federal timber sales). A study performed for the California Integrated Waste Management Board in 1990–91 examined cost-based disposal fees. The Tellus Institute, who performed the study, found that the conventional monetary costs of disposing of the State's waste were $3.5 billion, and estimated additional environmental costs, associated primarily with landfill leachate remediation, at $1.7 billion for an estimated "full disposal cost" in excess of $5 billion per year. Thus in this analysis, over a third of the estimated full cost of disposal was not reflected in actual disposal fees paid. Frank Ackerman, "Waste Management: Taxing the Trash Away," *Environment* (June 1992), 2–5.

Chapter 3

1. William Osborn, *The Paper Plantation* (New York: Grossman Publishers, 1974), 37.

2. Osborn, particulate matter and sulfur oxides, 116, 124; kraft mill odors, 103–4; chemicals handling, 44; Androscoggin River, 57.

3. The discussion of pollution control is based on Robert Gottlieb and Maureen Smith, "The Pollution Control System: Themes and Frameworks"; Robert Gott-

lieb, Maureen Smith and Julie Roque, "By Air, Water and Land: The Media-Specific Approach to Toxics Policies"; and other chapters in Robert Gottlieb, ed., *Reducing Toxics: A New Approach to Policy and Industrial Decisionmaking* (Washington, D.C.: Island Press, 1994).

4. Ronald Slinn, "The Paper Industry and Capital," *Tappi Journal* (September 1992), 16.

5. U.S. Environmental Protection Agency, Office of Air Quality Planning and Standards, *National Air Pollutant Emission Estimates 1940–1989,* EPA Publication no. EPA-450/4–91-004 (Research Triangle Park, N.C.: U.S. Environmental Protection Agency, March 1991).

6. U.S. Environmental Protection Agency, *Toxics in the Community: 1989 National and Local Perspectives* (Washington, D.C.: U.S. Government Printing Office, 1990); 1989 TRI Data (tape distribution from U.S. Government Printing Office).

7. David Treadwell, "Jobs vs. Pollution: Two States Clash Over River Cleanup," *Los Angeles Times* (11 July 1989).

8. An overview is found in Thomas J. McDonough, "The ABCs of Conventional Technologies Related to Pulping and Bleaching," Proceedings, International Symposium on *Pollution Prevention in the Manufacture of Pulp and Paper: Opportunities and Barriers,* 18–20 August 1992, Washington, D.C., EPA-744R-93-002 (Washington, D.C.: U.S. Environmental Protection Agency, February 1993) (hereafter PP Symposium), 35–40. Standard technical references for pulping technology are: *Pulp and Paper Manufacture,* 2d ed., a multivolume series published by the Joint Textbook Committee of the Pulp and Paper Industry (1983-); and James P. Casey, *Pulp and Paper: Chemistry and Chemical Technology,* 4 vols. (New York: John Wiley and Sons, 1983).

9. Bruce Fleming, "Alternative and Emerging Nonkraft Pulping Technologies," in PP Symposium, 78.

10. Alfred Wong and Jiri Tichy, "A Novel Low-Pollution Approach for the Manufacture of Bleached Hardwood Pulp," Proceedings, International Conference on Pollution Prevention, Washington, D.C., 10–13 June 1990; Al Wong, "Sulfite, the New Wave," preprint, PapFor 92 Conference, St. Petersburg, Russia, 20–23 September 1992 (Atlanta: TAPPI Press, 1992).

11. Wong and Tichy.

12. On anthraquinone see Wong and Tichy; "Research Team Seeks to Derive Pulping Catalyst from Lignin," *American Papermaker* (November 1991), 24; Michael Ringley "Westvaco Uses Anthraquinone to Increase Alkaline Pulp Yields," *American Papermaker* (April 1991), 52; and Robert Kinstrey, "Industry Searches for New Ways to Replace Chlorine Bleaching Agents," *American Papermaker* (July 1991), 48.

13. On organosolve pulping see Fleming, "Alternative and Emerging"; Jerome Koncel, "Alcell Pulping Process Moves to First Commercial Application," *American Papermaker* (January 1991), 22; Scott Jamieson, "Alcell Pulping: World Class

Research Right Here in Canada," *Pulp and Paper Canada* 92(3) (1991), 16; and Robert Kinstrey, "An Overview of Strategies for Reducing the Environmental Impact of Bleach-Plant Effluents," *Tappi Journal* (March 1993) (organocell process). On silvichemicals see U.S. Dept. of Agriculture, Forest Service, *An Analysis of the Timber Situation in the United States: 1989–2040,* General Technical Report RM-199, Fort Collins, Colorado (Washington, D.C.: U.S. Dept. of Agriculture, 1990).

14. On xylanase see Lubomir Jurasek and Michael Paice, "Saving Bleaching Chemicals and Minimizing Pollution with Xylanase," PP Symposium, 105–7.

15. Newton, "Pulp, Paper, and Presswood," *North Coast Journal* (June 1990), 3.

16. U.N. Food and Agriculture Organization (FAO), *Pulp, Paper, and Paperboard Capacity Survey, 1993–1998* (FAO, 1994); and *Lockwood-Post's Directory of the Pulp, Paper and Allied Trades* (San Francisco: Miller Freeman, 1995).

17. On API testing see Peter Von Stackelberg, "White Wash: The Dioxin Coverup," *Greenpeace* (March/April 1990), 10. The EPA has been formulating new regulations aimed at reducing dioxin and furan contamination by pulp mills: the result of a legal challenge under TSCA filed by the Environmental Defense Fund and the National Wildlife Federation. See Conservatree Paper Company, *ESP News* (March 1991), 6; and Terry Bass, "EPA Rule Will Limit Land Application of Sludge Containing Dioxin and Furan," *American Papermaker* (April 1992), 40.

18. An overview of organochlorine pollution from bleached kraft mills that takes a position against an AOX standard is found in R. Berry et al., "The Effects of Recent Changes in Bleached Softwood Kraft Mill Technology on Organochlorine Emissions: An International Perspective," *Pulp and Paper Canada* 92(6) (1991), 43. A recent survey that incorporates an assessment of the cluster rule impacts is found in Thomas J. McDonough, "Recent Advances in Bleached Chemical Pulp Manufacturing Technology," *Tappi Journal* (March 1995), 55–61.

19. Doug Smock, "How Will the Toxics' Debate Affect Plastics?" *Plastics World* (November 1994), 28; Joel Bleifuss, "Have Your PVC, and Dioxin Too," *In These Times* (6 March 1995), 12; and Bette Hileman, Janice R. Long, and Elisabeth M. Kirschner, "Chlorine Industry Running Flat Out Despite Persistent Health Fears," *Chemical and Engineering News* (21 November 1994), 12–26.

20. U.S. District Judge Owen Panner, quoted by Von Stackelberg, 7. For other Greenpeace perspectives see Andrew Davis, "Erasing Dioxin From the American Paper Mill," *Business and Society Review* (1990); and Renata Kroese, *The Greenpeace Guide to Paper* (Amsterdam: Greenpeace International, 1990).

21. Wong, "Sulfite"; Wong and Tichy.

22. Cost estimates cited here are from: Neil McCubbin, "Costs and Benefits of Various Pollution Prevention Technologies in the Kraft Pulp Industry," PP Symposium, 172; and Frank Steffes and Ulf Germgard, "ECF, TCF Upgrade Choices Key on World Market, Environmental Forces," *Pulp and Paper* (June 1995),

83–92. For general summaries of strategies to reduce chlorine use in pulp bleaching see McDonough, "Recent Advances"; and Kinstrey, "An Overview."

23. For a survey of oxygen delignification systems see Andy Harrison, "O_2 Delig Matures into Key Process Segment at Modern N.A. Fiberlines," *Pulp and Paper* (November 1994), 55–71.

24. Steffes and Germgard.

25. Nils G. Johansson, Fredrick M. Clark, and David E. Fletcher, "Developing Technologies Open Door to Future Closure of Bleach Plant," *Pulp and Paper* (June 1995), 71–75.

26. Harold L. Hintz, "Conventional Pulp Bleaching at Westvaco," PP Symposium, 44.

27. Jim Young, "Union Camp Leads Ozone Pulping Drive in North American Mills," *Pulp and Paper* (September 1994), 69; William H. Trice, "Bleaching Papermaking Pulps with Oxygen and Ozone in a Commercial Installation," PP Symposium, 100–104.

28. "Ontario Banning Chlorine," *Paper* (March 1993), 9. On British Columbia and Howe Sound see Mary Walsh, "Canadian Mill's Quandary: Can It Make Money on Pollution-Free Paper?" *Los Angeles Times* (8 March 1992); and Jacki Cox, "North American Pulp Producers Move into Chlorine-Free Bleaching," *American Papermaker* (July 1992).

29. Roland Lövblad, "The Driving Force Behind TCF Kraft Pulp," *Paper Technology* (January/Febrary 1993), 13; Margaret Rainey, "A Chlorine-Free Paper Economy—Europe on the Verge," PP Symposium, 161–5.

30. Richard Paddock, "Surfers Force Pulp Mill to Halt Ocean Pollution," *Los Angeles Times* (10 September 1991); Jim Young, "Louisiana-Pacific's Samoa Mill Establishes TCF Production," *Pulp and Paper* (August 1993) (describes the toxicity testing requirement added to LP's wastewater permit); Anton F. Jaegel and Kirk A. Girard, "TCF Bleaching and the Louisiana-Pacific Corporation Samoa Pulp Mill," undated technical paper received from Louisiana-Pacific in June, 1995; and "Ketchikan Fined $6.1 million for Toxic Dumping," *Pulp and Paper* (May 1995), 25. In additional developments, a federal grand jury was expected to indict LP for alleged criminal violations at its OSB (oriented strand board) mill in Montrose, Colorado, including alleged falsification of air quality permit data. "Louisiana-Pacific under Federal Probe," *Pulp and Paper* (July 1995), 25.

31. "Congressman: Kick Out Chlorine," *Paper* (October 1993), 8.

32. "Late-Breaking News," *American Papermaker* (June 1992), 1.

33. For the "natural occurrence of organochlorines" argument see Bruce Fleming, "Organochlorines in Perspective," *Tappi Journal* (May 1995), 93–98.

34. See, for an example, Bruce Fleming, "Environmental Benefits of TCF Oversold," and W. Henson Moore "Environmental Standards: Technology vs Performance," both in *Tappi Journal* (May 1995), 88–89 (responding to Helge Eklund

address discussed below). Also see rhetoric in Ivan Amato, "The Crusade to Ban Chlorine," *Garbage* (Summer 1994), 30–39.

35. Keynote address by Eklund reproduced in Helge Eklund, "ECF vs TCF—A Time to Assess and a Time to Act," *Tappi Journal* (May 1995), 83–87.

36. The primary sources for the discussion based on the work of Ashford et al. (all of which present the basic definitions used above) are: Nicholas Ashford and George Heaton, Jr., "Regulation and Technological Innovation in the Chemical Industry," *Law and Contemporary Problems* 46(3) (1983), 109–57; Nicholas Ashford, Christine Ayers, and Robert Stone, "Using Regulation to Change the Market for Innovation," *Harvard Environmental Law Review* 9(2) (1985), 419–466; and Nicholas Ashford, "A Unified Technology-Based Strategy for Incorporating Concerns about Risks, Costs, and Equity in Setting National Environmental Priorities," paper presented at the Conference on Setting National Environmental Priorities, hosted by Resources for the Future, Annapolis, Maryland, 16–17 November 1992.

37. For a description of PAPRICAN see Susan Clites, "Pulp and Paper Research Institute of Canada Exemplifies Centralized Research," *Tappi Journal* (October 1992), 61. Also see Robert McGrath, "Should There Be Some Centralized Research in the U.S. Paper Industry?" *Tappi Journal* (October 1992).

38. Jim Young, "Improved Economy, New Products Boost Pulping/Bleaching Activity," *Pulp and Paper* (June 1994); and Hileman et al. (1994). The discussion of structural changes in chlor-alkali production and consumption is additionally based on: Allison Lucas, "U.S. Mills Set to Invest in Environmental Improvements," *Chemicalweek* (9 February 1994), 25–27; Michael Roberts, "Structural Changes in Chlor-Alkali" and "Chlor-Alkali Makers Resolve to Fight Back and Boost Research," *Chemicalweek* (4 November 1992), 36, 31; Bruce Fleming, "Environmental Pressures Produce Chlorine and Caustic Imbalance," *American Papermaker* (February 1991), 48–49; Joe Piccione, "Changes in Bleaching and Papermaking Having Major Effect on Chemical Use," *Pulp and Paper Canada* 92(1) (1991), 15–22; Carl Verbanic "Can Chlorine and Caustic Recycle the Good Times?" *Chemical Business* (September 1990) (special advertising supplement), 23–35; Earl Anderson, "Sagging Chlorine Use Crimps Caustic Soda Supply," *Chemical and Engineering News* (21 May 1990), 19–20; and "Bleachers Battle to Brighten Paper," *Chemical Business* (April 1990), 31–32.

39. All are located near trona mines in southwest Wyoming. Trona ore is the source of much of the soda ash produced in the United States. Randall H. Shearin, "Elf Atochem Reopens Chemical Caustic Soda Plant," *Tappi Journal* (April 1995), 48.

40. Hileman et al., (1994), 16.

41. Fleming, "Environmental Pressures," 48.

42. Roberts (1992), 36.

43. Fleming, "Environmental Pressures," 48.

44. "Bleachers Battle to Brighten Paper," *Chemical Business* (April 1990), 31–32.

45. Verbanic, 30.

46. The chlorate sector of the European chemical industry commissioned a report in 1993, which concluded that "ECF and TCF are equally low in environmental impact." It became the basis of an information campaign by Eka Nobel, a major producer of sodium chlorate. Michael Roberts, "Pell-Mell Rush to TCF May Slow Down in Europe," *Chemicalweek* (9 February 1994), 28.

47. Lövblad, 13.

48. On Sweden see, for example, Don Hinrichsen, "Integrated Permitting and Inspection in Sweden," in Nigel Haigh and Frances Irwin, eds., *Integrated Pollution Control in Europe and North America* (Baltimore: The Conservation Foundation, 1990).

49. The "precautionary principle"—acting to prevent anticipated harm, even when it cannot yet be fully characterized—is an increasingly prominent concept in international environmental debates. It is strongly associated with the Nordic countries, where it was first endorsed by the North Sea conference in 1987, although its roots can be found in both U.S. and other national and international policies going back several decades. Daniel Bodansky offers a thoughtful overview of the difficulty of applying the precautionary principle to real world choices in "Scientific Uncertainty and the Precautionary Principle," *Environment* (September 1991), 4–5, 43–44. However, if there were ever a case in which precautionary and preventative approaches were warranted, the vast network of serious implications associated with chlorine use make it a good candidate. Also, see additional comments in text on combining precaution with flexibility.

50. Quoted in Lövblad, 14.

51. Quoted in Lövblad, 15. A summary of other European activities around the subject of organochlorine pollution from pulp mills can be found in: Jens Folke, "A Consultant's View of European Government Activities," PP Symposium, 209–215; and International Institute for Environment and Development (IIED), *The Sustainable Paper Cycle,* phase I review report (London: IIED, 1995), annex E.

52. Ashford, 21.

53. Ashford and Heaton, 120.

54. Mark Weintraub, Hans Grüfeld, and Pieter Winsemius, "Strategic Planning Cuts Green Costs," *Pulp and Paper International* (May 1995), 45.

55. The question of whether there is or is not a market demand for TCF pulps has been widely debated. In particular, see (in PP Symposium): David Assman, "The Market for Chlorine-Free Paper," 240–243; and Rainey. Georgia-Pacific argues that those pursuing TCF agendas are soley motivated by commercial interest because TCF mandates "would make today's high-cost TCF mills the low-cost producers, while turning currently low-cost mills into high-cost produc-

ers" in Clifford Howlett Jr., "The Right Balance—Environmental Responsibility and the Competitive Edge," 155–160. Similarly, Jens Folke has argued that the German interest in TCF paper is driven by the fact that most of its chemical pulp mills are sulfite mills, and, "The only competitive edge for a sulfite mill is that it's an easier pulp to bleach. Thus the Germans were interested in getting the AOX level down to zero because that would increase their competitive edge." Folke, 149. Also see comments by Fleming, "Environmental Benefits." Louisiana-Pacific, the only TCF kraft producer in the United States, is reported to be experiencing weak demand for its pulp. Recently, the company began pursuing legislation in California to establish a TCF purchasing preference to supplement the recycled-content purchasing preference in state paper procurement guidelines (A.B. 826, introduced by California State Assembly members Sher and Hauser, as amended 5 April 1995). Others argue that assessing TCF demand presents a chicken-and-egg problem (it is difficult to demonstrate strong market demand for products that mostly aren't on the market) and that paper producers are constrained from expanding TCF paper production by limited supplies and choices of TCF pulps, Eklund, 84.

56. Figure of $100 million from Lucas (1994); statements by Georgia-Pacific in Howlett; "twenty-year investment" statement in PP Symposium, 109.

57. PP Symposium, 169. The substance of the response by Clifford Howlett of Georgia-Pacific was that the dioxin cases were related to the litigious nature of U.S. society and that the suits mentioned were on appeal.

58. On the formation of the EPA pulp and paper cluster group see: Martha Prothro, "The Pulp and Paper Cluster's Mission," Mark Greenwood, "The Pulp and Paper Sludge Rule," and John Seitz, "The Maximum Achievable Control Technology Rule," all in PP Symposium.

59. Various materials distributed by the EPA on the MACT/effluent guideline regulatory alternatives approach meeting held 24 February 1993, in Durham, North Carolina.

60. Bill Nichols, "EPA's Proposed Cluster Rules Shape U.S. Paper Industry's Near Future," *Pulp and Paper* (September 1994), 75–85. Other sources for this discussion include: McDonough, "Recent Advances"; Subhash Chandra, "Implementation of EPA's BMP Plans Could Help Reduce Regulatory Load," *Pulp and Paper* (September 1994), 89–91; Yogesh M. Mehta, "Unbleached Mills Won't Escape Impact of EPA's 'Cluster Rules,'" *Pulp and Paper* (May 1995), 61–65; and Robert J. Crawford et al., "Emissions of Volatile Organic Compounds and Hazardous Air Pollutant from Oxygen Delignification Systems," *Tappi Journal* (May 1995), 81–91.

61. Nichols; and Weintraub et al.

62. Nichols, 81–82, raises the personnel availability question.

63. Frances Irwin, "Introduction to Integrated Pollution Control," in Nigel Haigh and Frances Irwin, eds., *Integrated Pollution Control in Europe and North America* (Baltimore: The Conservation Foundation, 1990), 10.

64. U.S. Environmental Protection Agency, Office of Pollution Prevention and Toxics, *Pollution Prevention News* (January/February and August/September 1994). The industries selected are: auto, computers and electronics, iron and steel, metal finishing and plating, petroleum refining, and printing.

65. G. H. Nelson, T. F. Clark, I. A. Wolff, and Quentin Jones, "A Search for New Fiber Crops: Analytical Evaluations" *Tappi* (January 1966), 40.

66. Lignin, ash, and silica content are based on the following (where they are in disagreement the maximum and minimum figures found are used for ranges): multiple sources as cited in Al Wong, "New Direction in Industry Development and Environmental Protection for Non-Wood Pulp Mills in Developing Countries," Proceedings, Second International Non-Wood Fiber Pulping and Papermaking Conference, Shangai, China, 6–9 April 1992; Frans Zomers, Richard Gosselink, Jan van Dam, and Bôke Tjeerdsma, "Organosolv Pulping and Test Paper Characterization of Fiber Hemp," *Tappi Journal* (May 1995), 149–55 (hemp only); various sources as cited in International Institute for Environment and Development (IIED), *Non Wood Fiber Substudy,* first draft (London: IIED, 1995), 4; Judt (1993, 1994), Van der Werf et al.(1994), Hurter (1995), Touzinsky (1993), all as cited in Paper Task Force, "Nonwood Plant Fibers As Alternative Fiber Sources for Making Paper," White Paper 13 (New York: Environmental Defense Fund, 1996), table 7; Northern Territory of Australia, Dept. of Primary Industry and Fisheries (1989), UNEP (1981), Mittal (1994), and Kilpinen (1992), all as cited by ReThink Paper (a project of Earth Island Institute), Kenaf Monograph, unpublished draft (December 1995).

67. Harmohinder Sabharwal et al., "Bio-Refiner Mechanical Pulping of Bast Type Fibers," Proceedings, 1994 TAPPI Pulping Conference, San Diego, California, 6–10 November, 1994.

68. Wong, "Technical and Economic Obstacles Affecting the Early Commercialization of Kenaf Pulp Manufacture in Developed Countries," Proceedings, 1991 TAPPI Pulping Conference, vol. 1, Orlando, Florida, 3–7 November 1991 (Atlanta: TAPPI Press, 1991), 506–7 (comparative flexibility of sulfite in pulping mixed fiber types); Wong, "New Direction," 30, 36, 38. On sisal he cites: N.M. Da Silva and A.D. Pereira, "Experience of a Pioneer—Sisal," Proceedings, TAPPI Pulping Conference, San Francisco, California, 1984.

69. Personal communication with Tom Rymsza, KP Products, Albuquerque, New Mexico (December 1995).

70. Joseph E. Atchison, "Nonwood Fiber Could Play Major Role in Future U.S. Papermaking Furnishes," *Pulp and Paper* (July 1995), 126.

71. Wong, "New Direction," 33.

72. Zomers et al.

73. Personal communication with Al Wong, Arbokem Ltd., Vancouver, British Columbia (August 1995); Canadian Flax Pulp, Ltd., *Environmental Prospectus for Proposed Wheat Fibre Plant in Vulcan, Alberta* (February 1994) (submitted and approved under the Alberta environmental assessment process).

74. Wong, "New Direction," 35–36, cites F. Li et al., "Disposal of Ammonium-Based Pulping Black Liquor of Small-Sized Pulp and Paper Mills as Practiced in the People's Republic of China," NIEM project phase 2 report (Beijing: Institute of Light Industry, Dept. of Chemical Engineering, 1990).

75. Personal communication with Colin Felton, U.S. Department of Agriculture, Forest Service, Forest Products Laboratory, Madison, Wisconsin (November 1995).

76. See, for example, "New Mills and Major Expansions," *Lockwood-Post's Directory* (especially 1992–95), and various announcements in *Pulp and Paper* and *Tappi Journal* during the same period. The Weyerhaeuser mill in Plymouth, North Carolina and the Potlatch mill in Cloquet, Minnesota are two that have substantially expanded capacity in the process of major environmental compliance upgrades.

77. In almost identical letters submitted under different names to the editors of the *Washington Post* and the *San Francisco Examiner* (in response to stories on kenaf in those papers), two executives of the AFPA offered imaginative (albeit unsubstantiated) new theories about waste disposal problems that might be associated with kenaf use. They then highlighted the "billions" of dollars being invested in deinking to provide future pulp sources, solve waste problems, and magnify the effective yield of trees, implying that kenaf use in papermaking would provoke a trade-off of these benefits. Each also argued that "millions of acres" of land would have to be cleared to grow kenaf for use in paper, which "could mean leveling forests" (Van Hook), "forests would have to be leveled" (Moore). However, most analysts agree that industrial fiber crops like kenaf compete with other uses (such as tree plantations) for marginal agricultural land and not with established forests. Letters by W. Henson Moore, President and CEO, AFPA, *Washington Post* (20 December 1995), and Matt Van Hook, Vice President, AFPA, *San Francisco Examiner* (9 March 1995).

78. A broader agenda could maintain the necessary focus on the issue, chlorine, but more overtly endorse preferred ways of getting there. This could emphasize alternative fiber use (both wastepaper and nonwoods) and alternative pulping technologies, and seek to better illuminate the linkage between these opportunities and the goal of eliminating the use of chlorine and chlorine compounds in pulp production.

79. For example, the EDF-based Paper Task Force's final report on guidelines for purchasing environmentally preferable paper generally provided positive reinforcement for recycling and reducing paper use, and an overview of how to examine the range of environmental impacts associated with paper production and use. Yet, on the critical issues of TCF versus ECF, production scale, alternative pulping technologies, and nonwoods, it essentially endorsed positions already reflected in existing regulatory and industry trade association programs. Within a few weeks of the report's release it had already been cited several times by industry groups as support for ECF over TCF production. This included the use of the report by industry opposition (Weyerhaeuser, AFPA, and International

Paper) in testimony on California Assembly Bill 826, which sought to incorporate a TCF preference into state procurement policy. The Task Force report was also endorsed by AET, "an international association of chemical manufacturers and forest products companies . . . created to establish a clearinghouse of educational and technical resources relating to chlorine dioxide and its use in chemical pulp bleaching." Letter from Emily Miggins to Lauren Blum (12 January 1996); AET web page at http://aet.org/new/taskforce.html (January 1996); and Paper Task Force, *Paper Task Force Recommendations for Purchasing and Using Environmentally Preferable Paper* (New York: Environmental Defense Fund, 1995).

80. C. A. Moore, W. K. Trotter, R. S. Corkern, and M. O. Bagsby, "Economic Potential of Kenaf Production," *Tappi* (January 1976), 117.

81. Jerry Powell, "Fewer Trees in AFPA's Forest," *Resource Recycling* (November 1995), 6.

82. W. Henson Moore, "The Paper Industry's New Outlook on Regulatory Reform," *Tappi Journal* (April 1995), 12.

83. Stefan Kay, "Pulp Prices That We Can Ill Afford," *Pulp and Paper International* (1994).

84. Weintraub et al., 46–47.

Chapter 4

1. Franklin Associates, Ltd., *Characterization of Municipal Solid Waste in the United States 1960–2000* (U.S. Environmental Protection Agency, Office of Solid Waste, PB-178323, 1986) (with updates in even-numbered years, most recently 1994).

2. Andrew Szasz, "In Praise of Policy Luddism: Some Strategic Lessons from the Hazardous Waste Wars" (unpublished draft, University of California, Santa Cruz, 1989). Szasz quotes extensively from the legislative hearings record in documenting his case. The following testimony of chemical manufacturers that he cites is representative of the form that industry opposition took:

[F]ree market economics should be the primary force for stimulating recovery or recycling of materials . . . We also object to the absolute blanket authority to control production processes and product composition. (Dow)

We believe that the disposal of wastes ought to be regulated instead of regulating the nature and use of the product or the type of manufacturing process used. (Dupont)

[L]egislation should not impede the natural interaction of raw materials, market and other forces that ultimately control the nature, quality, price, and success of products developed in our free enterprise system. (Union Carbide)

Solid Waste Disposal Act Extension—1974; Hearings before the Subcommittee on Public Health and Environment of the House Committee on Interstate and Foreign Commerce, 93rd cong., 2d sess. (28–29 May 1974), 35, 291, 292

(Democrats bill and Dow); and *The Need for a National Materials Policy; Hearings before the Subcommittee on Environmental Pollution of the Senate Committee on Public Works,* 93rd cong., 2d sess. (11–13 June, 9–11 and 15–18 July 1974) parts 2:154, 3:1748 (Dupont and Union Carbide), as cited in Szasz.

3. J. Quarles, *Federal Regulation of Hazardous Wastes: A Guide to RCRA* (Washington, D.C.: Environmental Law Institute, 1982), 3.

4. Midwest Research Institute, *Paper Recycling: The Art of the Possible 1970–1985* (Kansas City, Mo.: American Paper Institute, 1973), 176 (a report prepared by William Franklin, Robert Hunt, and Sally Sharp for the Solid Waste Council of the Paper Industry). This report provides an interesting benchmark for the reports later prepared by Franklin Associates for the American Paper Institute, *Paper Recycling: The View to 1995,* summary report (New York: API, 1990), and for its successor, the American Forest and Paper Association, *The Outlook for Paper Recovery to the Year 2000,* executive summary (Washington, D.C.: AFPA, 1993). The latter two reports are hereafter cited as Franklin/API (1990) and Franklin/AFPA (1993).

5. California Integrated Waste Management Board, Interim Database Project, Estimated Average 1990 Waste Stream Composition, revised 15 January 1993. The difference could have been accounted for by problems with the first implementation of the state reporting system (since substantially dismantled) and with the underlying assumptions made by cities and counties in auditing actual wastestreams, but they could also reflect state characteristics, such as the concentration of population in metropolitan areas, a large service sector in the economy, or wastepaper related to significant levels of international trade.

6. Göttsching estimated that 12 percent of paper consumption is destroyed or too contaminated for recovery and that another 6 percent is marked for long-term uses. Lothar Göttsching, "Issues and Opportunities in Wastepaper Recycling," in Gerard Schreuder, ed., *Global Issues and Outlook in Pulp and Paper* (Seattle: University of Washington Press, 1988), figure 2, 213. In modeling wastepaper recycling scenarios, the USDA has used estimated maximum recovery rates that range from 60–80 percent. Peter J. Ince and Dali Zhang, "Impacts of Recycling Technology on North American Fiber Supply and Competitiveness," Proceedings, Third International Pulp and Paper Symposium, University of Washington, Center for International Trade in Forest Products (CINTRAFOR), Seattle, 13–14 September 1994, 132–145 (hereafter CINTRAFOR Proceedings).

Jaakko Pöyry, the largest paper industry consulting firm in the world, continues to estimate the maximum practical recovery rate at a low 48–50 percent as referenced by Tara Kern, "Recovered Paper Markets: What Does the Future Look Like?" *Resource Recycling* (November 1995), 48–56. However, the following countries have all been identified as having achieved recovery rates above 50 percent by 1993: Austria (68.4 percent), The Netherlands (62.7), Switzerland (54.4), Germany (54.2), Japan (51.1), and Sweden (50.4). Mahendra Doshi, Gary Scott, and John Borchardt, "Semiannual Conference Review: January-June 1995," *Progress in Paper Recycling* (August 1995), 81.

7. U.S. Congress, Office of Technology Assessment, *Facing America's Trash: What Next for Municipal Solid Waste?* (Washington, D.C.: OTA, 1989), 365.

8. "Paper Recycling Approaches," *BioCycle* (October 1992), 46–53. The article reported the API denominator as "tons paper produced," but the numbers and API (now AFPA) publications (e.g., Franklin/AFPA, 1993) do not support this. The API/AFPA has consistently used a recovery rate based on apparent consumption, which they call "new supply."

9. Paper Task Force, *Paper Task Force Recommendations for Purchasing and Using Environmentally Preferable Paper* (New York: Environmental Defense Fund, 1995), chapter 3. For the underlying technical studies see Paper Task Force, *Virgin and Recycled Paper Manufacturing,* technical supplement, part V (New York: Environmental Defense Fund, 1995).

10. William Kovacs, "The Coming Era of Conservation and Industrial Utilization of Recyclable Materials," *Ecology Law Quarterly* 15 (1988), 551.

11. U.N. Food and Agriculture Organization (FAO), Wood Industries Branch, *Paper Recycling Scenarios* (FAO, 1994) (prepared by Jaakko Pöyry Consulting AB, Sweden).

12. For a discussion of the drawbacks of using life-cycle analysis in the comparative evaluation of paper recycling scenarios see International Institute for Environment and Development (IIED), *Towards a Sustainable Paper Cycle,* final report review draft (London: IIED, 1996), 178–82.

13. Table 4.1 is based on data reported in Franklin/API (1990) for 1970–88; Franklin/AFPA (1993) for 1989–92; and *Lockwood-Post's Directory of the Pulp, Paper, and Allied Trades* (San Francisco: Miller Freeman, 1995) for 1993–94. Lockwood-Post's data is generally highly consistent, although not always an exact match, with the Franklin/trade association data for production, trade, and wastepaper use.

14. Anita Menniga, "Paper Recovery Begins at Home," *Pulp and Paper International* (1994), 5.

15. California Integrated Waste Management Board, *1996 Market Development Plan,* staff draft (Sacramento: CIWMB, 1996), appendix 6.

16. The data on which both figure 4.1 and table 4.2 are based come from different sources that are not completely consistent. Only USDA data provides the full set of figures necessary to calculate wood fiber utilization by different fiber sources, but it is typically 2–5 years out of date by the time the complete set of figures becomes available. Generally, earlier data is from the USDA, and later data is from trade sources. Data used also include the author's calculations and assumptions as described below, and are thus rough approximations. However, although specific figures (e.g., for total wastepaper used in any year) may differ substantially by source, the general trends depicted in the aggregated data are consistent with trends within each source. As far as possible, a single source (or consistent sources) were used for the periods in which major adjustments

were occurring (e.g., 1940–1960 and 1980–1994). Specific figures were derived from the following major sources as described further below:

USDA1 U.S. Dept. of Agriculture, Forest Service, *An Analysis of the Timber Situation in the United States: 1952–2030,* forest resource report no. 23 (Washington, D.C.: USDA, 1982).

USDA2 U.S. Dept. of Agriculture, Forest Service, *An Analysis of the Timber Situation in the United States: 1989–2040,,* general technical report RM-199, Fort Collins, Colorado (Washington, D.C.: USDA, 1990).

DOC U.S. Dept. of Commerce, Bureau of Economic Analysis, *Business Statistics, 1963–91* (Washington, D.C.: U.S. Dept. of Commerce, Bureau of Economic Analysis, 1992).

FA1 Franklin Associates, Ltd., *Paper Recycling: The View to 1995,* summary report (New York: American Paper Institute, 1990).

FA2 Franklin Associates, Ltd. *The Outlook for Paper Recovery to the Year 2000,* executive summary (Washington, D.C.: American Forest and Paper Association, 1993).

MF1 *Lockwood-Post's Directory of the Pulp, Paper and Allied Trades* (San Francisco: Miller Freeman, various annual editions).

MF2 *Pulp and Paper 1993 North American Factbook* (San Francisco: Miller Freeman, 1994).

FAO U.N. Food and Agriculture Organization (FAO), *Forest Products Yearbook* (Rome: FAO, 1991).

A. Total (actual) woodpulp consumption
1920–79: USDA1, 298; 1980–85: USDA2, 33, table 27; 1986–94: MF1.

B. Net woodpulp imports, based on woodpulp exports and imports
1920–79: USDA1, 299; 1980–83: USDA2, 33, table 28; 1987–94: MF1.

C1. Woodpulp (residue)

C2. Woodpulp (roundwood)
1920–79: USDA1, 300; 1980–86: USDA2, 34. Based on the following:

pr = pulpwood produced (roundwood)
pc = pulpwood produced (plant by-products
pe = pulpwood exports
pi = pulpwood imports

where:

x = (pc – pe + pi) / pr
C1 = (A – B) (x)
C2 = (A – B) (1 – x)

These estimates (C1, C2) of the proportion of residue and roundwood in domestic woodpulp production (and consumption) further reflect the following assumptions:

(1) That all international pulpwood trade is chips from whole trees. In reality, pulpwood trade includes: roundwood and split wood, chips from whole trees, and wood residue (e.g., from sawmills). However, FAO data (1980–91) indicates that typically more than 95 percent of pulpwood exports are chips (Lockwood-

Post data indicates zero exports of roundwood pulpwood after 1989), about five percent are roundwood or split wood, and none is wood residue. Pulpwood imports have a higher proportion of roundwood and wood residue, but total pulpwood imports are small compared to exports.
(2) That *net* (apparent) pulpwood chip consumption in the United States is all from wood residue, and that all pulpwood roundwood production in the United States is consumed domestically (i.e., none is exported). Thus, net pulpwood chip consumption, divided by total pulpwood roundwood production, provides an estimate of the proportion of residue to primary wood in domestic woodpulp production and consequently in consumption of domestically produced woodpulp. The basic correlation is generally supported by the FAO data, also by the fact that USDA1 identified all pulpwood chips produced (through 1979) as "plant by-products." U.S. data do not formally distinguish between chips from whole trees, and chips from residue (see USDA2, 36); however, available industry and USDA estimates of the relative proportions of residue and primary pulpwood consumed are consistent with estimates as calculated here (see USDA2, 197, and Ronald Slinn, "Fiber from Forests: Facts and Figures," *Tappi Journal* (December 1992), 14).
(3) The ratio (x) is arbitrarily assumed to be 0.4 for 1987–94 (consistent with previous trends as calculated and with Slinn).

D. Total wastepaper consumption (by mills)
1920–69: USDA1, 298; 1970–92: average of USDA2, MF1, MF2, DOC, FA1, FA2; 1993–94: MF1.

Data for wastepaper use by domestic paper manufacturers are highly inconsistent between sources. In 1979, for example, the following figures were reported (in million tons): 12.9 (USDA1), 13.0 (USDA2), 15.3 (FA1), 14.3 (DOC), 15.4 (MF1, MF2). The average of reported figures, however, is not significantly inconsistent with rate of growth or actual consumption as reported in any one series. An average is used here for the years 1970–92 to minimize the disjuncture with USDA wastepaper use data (the only source available prior to 1970) and to permit a depiction of the long-term historical trends. (Not all series contain figures for all years between 1970–92.) In other tables (e.g., table 4.1) limited to more recent time frames, USDA data is not used and the figures (for wastepaper used and wastepaper utilization rate) will differ slightly from those calculated here from averages.

E. Total "other fiber" consumption
1920–79: USDA1, 300; 1980–86: USDA2, 34; 1987–94: MF1.

F. Total fiber consumption
Where:

F = B + C1 + C2 + D + E

This figure preserves the "apples-and-oranges" convention of showing *woodpulp* consumption (distinct from *pulpwood* consumption) as analogous to wastepaper and other fiber consumption. Because wood has a pulp yield of roughly 40–95 percent (averaging around 50 percent), this dramatically understates actual wood use relative to wastepaper and other fiber use. On the other hand, much of the total wood consumed by the industry is used for energy production, so the contribution of wood to *fiber content* in paper is appropriately reflected by woodpulp consumption, whereas the contribution of wastepaper used (with a pulp yield of roughly 70–95 percent) and other fiber used (with pulp yields of roughly 40–95 percent) to actual fiber content in paper is overstated. The total fiber utilization rate, which exceeds 100 percent, primarily reflects the excess nonfibrous content of wastepaper and other fiber that is lost (or used for energy production) in pulping those materials. The actual total fiber content of paper is less than 100 percent due to the use of fillers and coating in paper.

G. Total paper and paperboard production
1920–69: USDA1; 1970–83: FA1; 1984–94: MF1.

H. Utilization rates (UR)
Calculated by dividing fiber use by total paper and paperboard production, e.g.: Wastepaper UR = D / F.

17. "Paper Recycling Approaches," *BioCycle* (October 1992).

18. Recent trade in chips from *Lockwood-Post's Directory* (various years); other figures from sources as described for tables 4.1 and 4.2, and figure 4.1.

19. Clark Wiseman, "Increased U.S. Wastepaper Recycling: The Effect on Timber Demand," *Resources, Conservation and Recycling* (January 1993), 110.

20. Peter J. Ince, *Recycling and Long-Range Timber Outlook: Background Research Report, 1993 RPA Assessment Update, USDA Forest Service* (Madison, WI: U.S. Dept. of Agriculture, Forest Products Laboratory, 1994), 102. See also the interesting discussion of the background to the study at the beginning of the document.

21. Long-range forecasts based on the NAPAP model that incorporate various of these interventions are found in: Ince and Zhang; and Thomas C. Marcin, Irene A. Durbak, and Peter J. Ince, "Source Reduction Strategies and Technological Change Affecting Demand for Pulp and Paper in North America," CINTRAFOR Proceedings.

22. Colin C. Felton, Timber Demand and Technology Assessment, USDA Forest Products Division, "Economic and Environmental Assessments of Paper Manufacturing with Non-Forest Fibers," proposal submitted to National Research Institute, Competitive Grants Program (16 January 1996).

23. These observations are based on personal conversations and panel discussions at the Northern California Recycling Association annual conference (November 1994), the Tomales Bay Conference on wood conservation (October 1994), and the California Resource Recovery Association annual conference (June 1995).

24. Definitions here are based on the final rule issued by the EPA governing federal procurement of recycled paper products as mandated by RCRA: "Guideline for Federal Procurement of Paper and Paper Products Containing Recovered Materials," *Federal Register* (1988), 23,546–23,566; and in *Federal Register*

(1991), 49,992. The latest proposed updates to the guidelines were published in "Paper Products Recovered Materials Advisory Notice," *Federal Register* (15 March 1995), 14,182–14,191.

25. Figures are from Franklin/AFPA (1993), supplemented by production data from *Lockwood-Post's Directory* (1995).

26. This is slightly confusing because the primary pulp used for these paperboard grades is not the only pulp used. "Unbleached kraft" paperboard, for example, actually has a wastepaper utilization rate in excess of 15 percent (as of 1992), and "recycled" paperboard also contains virgin pulp. The EPA provided guidelines in its procurement standards (effectively for "*recycled* recycled paperboard"), which describe specific recycled paperboard products in terms of postconsumer material content (generally above 80 percent).

27. Ken L. Patrick, "Sharper Focus on Customer Boosts Industry's Interest in Mini-Mills," *Pulp and Paper* (December 1994), 75–79.

28. EPA, *Federal Register* (1988), 23,554.

29. TRI reports examined for 1988 and 1989 indicated high levels of both chlorine and chlorine-dioxide emissions from several of the handful of semi-integrated paper mills engaged at the time in deinking wastepaper for printing and writing-grade paper use. A Presidential Executive Order has since taken effect requiring postconsumer content, and new EPA guidelines have done the same.

30. Claudia Thompson cites Andover International Associates, *Projected U.S. Demand for Recycled Printing and Writing Paper* (1991) for an estimate that combined federal, state, and local government purchases of printing and writing paper amounted to 1.7 million tons (7 percent) of total annual consumption of these grades in 1990. Claudia Thompson, *Recycled Papers: The Essential Guide* (Cambridge, Mass.: MIT Press, 1992), 19. The EPA states that private sector purchasers "represent 95 percent or more of paper demand." *Federal Register* (15 March 1995), 14,183. Thus, "less than 3 percent" seems a safe assumption for total federal consumption of paper relative to total domestic consumption.

31. See Kovacs's (1988) classic accounting and assessment of the federal role in recycling development in the 1970s and 1980s upon which this discussion is based. Also see the discussion by Blumberg and Gottlieb, who establish a broader framework for viewing the federal role in solid waste developments. Louis Blumberg and Robert Gottlieb, *War on Waste: Can America Win Its Battle with Garbage* (Washington D.C.: Island Press, 1989), 60–67 and elsewhere.

32. Kovacs, 547–548.

33. For an overview of state efforts see Kovacs, 560–590 and elsewhere. On both the states, and in particular the role and history of incineration, see Blumberg and Gottlieb, 67–72, chapters 2 and 4–6.

34. *Outside* (April 1991), 12. By 1993, however, *Outside* was being printed on a "recycled paper" that is "bleached with oxygen."

35. Kovacs (1988), 551–55.

36. This discussion is based on two good analyses of the incinerator regulations: David Littell, "The Omission of Materials Separation Requirements from Air Standards For Municipal Waste Incinerators: EPA's Commitment to Recycling Up in Flames," *Harvard Environmental Law Review* 15 (1991), 601–635; and Karen Kendrick-Hands, "Clean Air Act Amendments of 1990," *Detroit College of Law Review* 1 (1991), 155–159.

37. Supreme Court decisions handed down in May 1994, however, required incinerator ash to be managed as a hazardous waste and also struck down local flow control ordinances as unconstitutional. Congressional efforts to restore some flow control powers were subsequently undertaken. Robert Steutville, "The State of Garbage in America," 35, and Dexter Ewel, "Solid Waste Flow Control Update," 38–39, in *Biocycle* (May 1995).

38. Sibbison, "Dan Quayle, Business's Backdoor Boy," *The Nation* (29 July 1991), 141.

39. Littell, 615.

40. Kendrick-Hands, 156–157.

41. Debra Garcia, "Recycling Capacity to Increase at Record Rates as Laws Proliferate," *Pulp and Paper* (May 1990), s1-s29.

42. For a summary of various state efforts to develop markets for secondary materials through the late 1980s, see John Ruston, "Developing Markets for Recycled Materials," (Environmental Defense Fund working paper, 1988). For a summary of recycled newsprint legislation by states through 1990 see Garcia.

43. Michael Alexander, "Developing Markets for Old Newspapers," *Resource Recycling* (July 1994), 20.

44. The new mills were: Alabama River Newsprint, Perdue Hill, Alabama (a joint venture of Atibiti-Price and Parsons and Whitmore); Newsprint South, Grenada, Mississippi; and Ponderay Newsprint, Usk, Washington (a joint venture of Canadian Pacific Forest Products and several newspaper publishers). Robert W. Dellinger, Virgil Horton, and Darlene Snow, "Waiting for the ONP Market to Improve," *Waste Age* (June 1990); and *Lockwood-Post's Directory* (1990–93).

45. Except where otherwise indicated, all figures and calculations in this section, including table 4.4, are based on figures reported in: U.S. Dept. of Commerce, Bureau of Economic Analysis, *Business Statistics, 1963–91* (Washington, D.C.: U.S. Dept. of Commerce, Bureau of Economic Analysis, June 1992) (1960s figures only); Franklin/API (1990) and Franklin/AFPA (1993), with production figures for the 1970s from *Lockwood-Post's Directory* (various years).

46. Midwest Research Center.

47. Regional capacity is from Garcia.

48. The *Register-Guard* (9 August 1990), C-1.

49. H. Mason Sizemore, president and chief operating officer of the *Seattle Times,* in a speech to the Canadian Business Outlook Conference (29 May 1990),

reprinted in: *Development of Recycling Markets: Hearings before the Subcommittee on Transportation and Hazardous Materials of the House Committee on Energy and Commerce,* 102d cong., 1st sess. (13 and 19 June 1991), 449–455.

50. Sizemore, 454.

51. Statement of Red Cavaney, President of the American Paper Institute, in *Development of Recycling Markets* (1991), 414.

52. Cavaney, and American Paper Institute, *Key Questions and Answers on Paper Recycling and its Role in Municipal Solid Waste Management* (brochure published by the American Paper Institute, 1990).

53. Statement of James Walden in *National Recyclable Commodities Act: Hearing before the Subcommittee on Commerce, Consumer Protection, and Competitiveness of the House Committee on Energy and Commerce,* 101st cong., 2d sess. (28 June 1990).

54. Blumberg and Gottlieb, 199–200.

55. Statement of James Walden, Paper Recycling Coalition, in *Development of Recycling Markets,* 474.

56. Jackie Cox, "Persistence, Common Sense Are the Qualities behind Ponderosa's Success," *American Papermaker* (June 1992), and Cox, "Shelby Tissue Is not Just Another 'Run of the Mill' Operation," *American Papermaker* (July 1992). In 1988, the most highly paid of major paper company executives was pegged by the Association of Western Pulp and Paper Workers as Michael Smurfit, CEO of Jefferson Smurfit, who received more than $9 million in total compensation. The second-place finisher was the CEO of Potlatch, who received nearly $5 million, followed by Georgia-Pacific's CEO, who received nearly $3.3 million. The median compensation of paper industry CEOs was about $830,000. James Thompson, "CEO Pay Escalates Higher," *The Rebel* (17 July 1989), 2.

57. For good, short summaries of these developments, see Eric Weltman, "Jobs Versus the Environment: No," *Public Citizen* (September/October 1993), 20–21; and Neil Seldman, "Recycling as Economic Development: We Can Invent Our Future," *Race, Poverty and the Environment* (Winter 1993).

58. Weltman.

Chapter 5

1. Nina Munk, "Self-Inflicted Wounds," *Forbes* (13 September 1993), 71, 74.

2. MacMillan Bloedel, Ltd., Environmental Assessment, Statement of Negative Declaration (filed with the City of West Sacramento, Community Development Department, 10 October 1991).

3. City of West Sacramento, *Revised Draft Supplemental Environmental Impact Report: MacMillan Bloedel Ltd. and Haindl Papier Recycled Waste Paper Facility, Southport Industrial Park* (May 1992) (prepared by Dames and Moore) (hereafter Revised Draft EIR), 22. Although Haindl was a partner in the proposal,

it was a silent one, so MacMillan Bloedel, represented in the permitting battles by one of its vice presidents and a local law firm, quickly became the focal point of attention.

4. The Coquitlam project (located in Coquitlam, British Columbia, within the greater Vancouver district) was an early case in which local forces combined to threaten a proposed deinking mill. The project, since completed and now operating, had a proposed capacity of about 350 tpd of recycled newsprint market pulp, which would be consumed by the province's two largest newsprint producers: MacMillan Bloedel and Fletcher Challenge Canada. Opposition arose in 1990 when a coalition of local environmental groups, joined by a political candidate, raised concerns about the toxicity of the sludge that would be produced and about where and how it would be disposed of. The Newstech mill, as it is known, was designed and operated by Hipp Engineering; the firm that also designed the West Sacramento mill. It was the first deinked market pulp mill in North America. See: Olivia Scott, "Save De-inking Plan," *The Province* (1 August 1990); Susan Combs, "British Columbia Moves Ahead on Stand-Alone De-inking Plant," *Recycling Times* (14 August 1990), 1; Katherine Monk, "Coquitlam Mayor Defends Proposed De-inking Plant," *Vancouver Sun* (1 August 1990); and Jackie Cox, "Western Canada's First Recycled Market Pulp Mill is Coming Online," *American Papermaker* (August 1991), 22–23.

5. City of West Sacramento, *Draft Supplemental Environmental Impact Report: MacMillan Bloedel Ltd. and Haindl Papier Recycled Waste Paper Facility, Southport Industrial Park* (January 1992) (prepared by Dames and Moore).

6. City of West Sacramento, *Final Supplemental Environmental Impact Report: MacMillan Bloedel Ltd. and Haindl Papier Recycled Waste Paper Facility, Southport Industrial Park* (August 1992) (prepared by Dames and Moore) (hereafter Final EIR), appendix C.

7. Final EIR, comment 28.

8. Letter from Norm Masters to Ken Selover, Yolo Solano Air Pollution Control District, 8 October 1992.

9. However, by way of illustrating existing models, MacMillan Bloedel took city officials on a junket to a rural deinking facility in Georgia, which became the subject of some disparagement in the Sacramento media. As one editorial observed:

> When a group of West Sacramento city leaders returned from [their] trip to Georgia . . . they gleefully reported that the plant emitted none of the foul odors normally associated with paper production facilities. It's nice to know that the proposed plant . . . betrayed no offensive stink. But the boosterism of city officials is nonetheless disconcerting, because it seems the critical eye they should be casting toward this project has stopped at their collective nose.

Sacramento News and Review (26 March 1992).

10. Revised Draft EIR, 87.

11. Air quality impacts discussed in Revised Draft EIR, 80–101.

12. Under regulatory standards for new sources of air pollution, emissions offset credits (or "pollution rights") would have to be purchased for each pollutant exceeding its new source action level. These credits could be purchased from existing stationary air pollution sources (e.g., manufacturers or energy utilities) who had reduced their emissions of the pollutant(s) from previously documented levels and could market the difference.

13. Planning Commission, City of West Sacramento, *Findings of Fact and Statement of Overriding Consideration for MacMillan Bloedel Ltd. and Haindl Papier Recycled Waste Paper Facility* (August 1992) (hereafter Statement of Overriding Consideration). Eventually, when permits for Phase III were sought (a decade or so down the line) the sum of emissions from Phases I, II, and III would finally trigger the new source requirements, and the permitting of Phase III would require the purchase of offset credits. In reality, Phase III would very likely be subject to a much stricter set of requirements as air pollution regulations evolved. Indeed, Phase II permits were being sought well in advance of construction most likely to avoid this problem; stricter action levels for new sources of pollution had already been defined, although they were not yet in effect. At a minimum, however, the segmentation strategy deferred the expense and the complications around the current unavailability of offset credits for NOx emissions for at least a decade, and justified the failure to complete a comprehensive health risk assessment for the project.

14. Final EIR, comment 2–6.

15. Letter from Leslie Krinsk to Harry Gibson, 12 August 1992. Reproduced as an attachment in Statement of Overriding Consideration.

16. Revised Draft EIR, 107.

17. Final EIR, comments 3–1 to 3–14 (Regional Water Quality Control Board) and 4–1 to 4–14 (Department of Water Resources).

18. Personal communication with Edward Means, Director of Water Quality, Metropolitan Water District, 17 August 1992.

19. Final EIR, comment 8–1.

20. Statement of Overriding Consideration, 128.

21. Personal communication with Norm Masters, West Sacramento (January 1993).

22. Final EIR, 3–70 (response to comment 18-03).

23. Wagonner v. City of West Sacramento, no. 70446 (Superior Court, filed 30 October 1992).

24. Revised Draft EIR, 166.

25. Letter from J. G. Kirkland to Harry Gibson, 12 August 1992. Reproduced as an attachment in Statement of Overriding Consideration.

26. Final EIR, comment 28. The company did not estimate the return on the project at completion of Phase III.

27. Statement of Overriding Consideration, 125.

28. Final EIR, 3–72 (response to comment 19-01).

29. Final EIR, 3–135 (response to comment 37).

30. City of West Sacramento, Planning Commission Meeting (4 June 1992), verbatim minutes.

31. Final EIR, 3–67 (response to comment 17-01).

32. Final EIR, comments 22–12 to 22–14 (QUAD comments) and pages 3–90 to 3–92 (response to QUAD comments).

33. Helen Maserati, "British Columbia Runs into Buzz Saw of Criticism on Plan to Log Rain Forest," *Los Angeles Times* (28 May 1993). The provincial government had purchased shares in MacMillan Bloedel only a few weeks before deciding to allow the company to log the Clayoquot Sound forest.

34. See, for example, California Integrated Waste Management Board, *1996 Market Development Plan,* staff draft (Sacramento: CIWMB, 1996). In this report as in others, the Board declined to offer any policy guidance on or even to express an interest in its own conclusions that wastepaper utilization capacity in the state has actually declined in recent years, that the most recent paper recycling mill was built more than ten years ago, and that the out-of-state wastepaper export rate is likely in the range of 55 percent or more and still climbing.

35. Thomas C. Marcin, Irene A. Durbak, and Peter J. Ince, "Source Reduction Strategies and Technological Change Affecting Demand for Pulp and Paper in North America," Proceedings, Third International Pulp and Paper Symposium, University of Washington, Center for International Trade in Forest Products, Seattle, 13–14 September 1994, 147–155.

36. Quoted in Munk.

Chapter 6

1. The economist Herman Daly has provided, most recently in his work with theologian John Cobb, the most stark illumination of the underlying irrationality of contemporary economic thought with respect to the environment. Herman E. Daly, *Steady-State Economics,* 2d ed. (Washington, D.C.: Island Press, 1991); Herman E. Daly and John B. Cobb, *For the Common Good: Redirecting the Economy toward Community, the Environment, and a Sustainable Future,* 2d ed. (Boston: Beacon Press, 1994). An accessible overview of major themes and conflicts in the development of economic thought vis à vis the environment and social justice, written in the context of the practical application of twentieth-century economic development theory is found in Benjamin Higgins and Jean Downing Higgins, *Economic Development of a Small Planet* (New York: W.W. Norton, 1979). These sources contain extensive references to the larger community of thinkers preoccupied with the environmental critique of dominant economic thought.

2. André Gorz, *Ecology as Politics* (Boston: South End Press, 1980).

3. For examples, see Donald F. Barnett, "The U.S. Steel Industry: Strategic Choices in a Basic Industry," in Donald A. Hicks, ed., *Is New Technology Enough: Making and Remaking Basic Industries* (Washington, D.C.: The American Enterprise Institute, 1988); Keith Chapman, *The International Petrochemical Industry: Evolution and Location* (Oxford: Basil Blackwell, 1991). Silicon Valley is widely cited for its value as a model of successful industrial agglomeration (and as a "geographically prestigious," "high-amenity" suburb). The latter terms are taken from James O. Wheeler and Peter O. Muller, *Economic Geography* (New York: John Wiley and Sons, 1986). For a rare alternative analysis, see Janice V. Mazurek, "How Fabulous Fablessness: Environmental Planning Implications of Economic Restructuring in the Silicon Valley Semiconductor Industry" (master's thesis, UCLA Dept. of Urban Planning, 1994). Mazurek documents and analyzes the high environmental premium paid for this otherwise stunning economic success.

4. Life-cycle inventories can provide a synopsis of available information, and the process of articulating the chains of relationships can help provide some conceptual clarity. However, the abstraction of the approach—which is blind to place, structure, people, policy, and (above all) change—combined with its dependence on bodies of quantitative, inevitably ad hoc, incomplete, and often contentious information (measures of environmental impact) make it a limited tool at best. Daly and Cobb's (1994) elaboration of "the fallacy of misplaced concreteness" in the discipline of economics is relevant to the critique of life-cycle analysis.

5. Annemieke Roobeek, *Beyond the Technology Race: An Analysis of Technology Policy in Seven Industrial Countries* (Amsterdam: Elsevier Science Publishers B.V., 1990), 221. The overview additionally relies on a series of accessible articles and essays reproduced in Vernon Whitford, ed., *American Industry* (New York: H.H. Wilson, 1984).

6. Congressional Budget Office, *The Industrial Policy Debate* (Washington, D.C.: U.S. Congress, 1983) and Robert Wescott, "U.S. Industrial Approaches: A Review," *Economic Impact* (August 1984), both as cited in Roobeek.

7. *Business Week* (4 July 1983) as reprinted in Whitford, 65.

8. Quotes from ibid., 56, 64.

9. From an article by Samuel Bowles, David M. Gordon, and Thomas E. Weisskopf in *The Nation* (June 1983) as reprinted in Whitford, 70.

10. Robert Gottlieb has explored the development of these mainstream and alternative tendencies and the ways they have interacted in the extended environmental community. He finds the roots of the alternative movements within the historical context of nineteenth- and early twentieth-century urban social movements, thus providing a groundbreaking expansion of the conventional domain of "the history of American environmentalism." Robert Gottlieb, *Forcing the Spring: The Transformation of the Environmental Movement* (Washington, D.C.: Island Press, 1993). Mark Dowie pursues a similar framework, offering additional detail on developments during the Clinton administration in *Losing Ground: American Environmentalism at the Close of the Twentieth Century* (Cambridge, Mass.: The MIT Press, 1995).

11. Project on Industrial Policy and the Environment, *America's Economic Future: Environmentalists Broaden the Industrial Policy Debate* (New York: Natural Resources Defense Council, 1984), 56–67.

12. The "technology vision" reflected in the Agenda was constructed by a technology working group of the American Forest and Paper Association, with the help of selected institutional workshop participants including the American Paper Machinery Association, the Technical Association of the Pulp and Paper Industry, other industry groups and various research centers. American Forest and Paper Association, *Agenda 2020: A Technology Vision and Research Agenda for America's Forest, Wood and Paper Industry* (Washington, D.C.: AFPA, 1994), 22–23. Absent from the process of devising an agenda to be funded by a first-year commitment of $45 million in public funds were, among others, the national and grassroots environmental groups who had been centrally included in the EPA's Common Sense Initiative and instrumental in the rise of public pressure for change in the industry.

13. For a powerful analysis of the central traditions in public planning see the classic work by John Friedman, *Planning in the Public Domain: From Knowledge to Action* (Princeton, N.J.: Princeton University Press, 1987). Although deeply grounded in theory, the book also offers accessible practical guidance on public planning processes. Those seriously intrigued by the prospects for pursuing focused democratic planning approaches to sustainable development may find Friedman's thoughts on the "recovery of political community" and "the mediations of radical planning" (chapters 9 and 10) both interesting and motivating.

14. A few examples (limited to efforts focused at local and regional levels) are: efforts led by cities such as San Jose, California, and state organizations such as the Clean Hawaii Center to evaluate and pursue different types of paper industry development that might be drawn into an area; the coordination between environmental groups, public health organizations, local economic development interests, and farmers in California reflected in the California Agripulp Mill Working Group; various community organizing efforts centered around particular mills, such as those focused around organochlorine pollution from the Stone Container mill in Missoula, Montana, led by the local Coalition for Health, Environment, and Economic Rights (CHEER), and around the Louisiana-Pacific mill in Ketchikan, Alaska; planning efforts undertaken within the context of a university/community project involving teams of student, labor, and community leaders focused partly on paper converters in a low income area of Los Angeles County; and collaborative efforts between the Natural Resources Defense Council, a community group, and various paper companies to finance and build a recycling mill in the South Bronx of New York. San Jose: various materials from the City of San Jose, 1994; Hawaii: various materials from the Clean Hawaii Center, and the Hawaii Dept. of Business, Economic Development, and Tourism (Honolulu), and personal communication with Mary Hieb, MLH Associates (Maui), October 1995; California: personal communication with Oakland Recycling Association, Materials for the Future Foundation, Earth Island Institute, and other organizations based in Oakland and San Francisco, 1994–1996; Mon-

tana: *Everyone's Backyard* (Summer 1995); Alaska: personal communication with Elsan Zimmerly (Ketchikan), December 1995; Los Angeles: personal communication with Robert Gottlieb (Los Angeles), 1995, and Carlos Porras (Los Angeles), February 1996; New York: John Holusha, "Paper Recycling Plant Planned for South Bronx," *New York Times* (5 June 1994); Weld Royal, "Paper Mill Project Plants Roots in the South Bronx," *BioCycle* (July 1994), 48–50; and Lis Harris, "Banana Kelly's Toughest Fight," *New Yorker* (24 July 1995), 32–40.

15. Robert M. Kaus, "Can Creeping Socialism Cure Creaking Capitalism?" *Harpers* (Fall 1983), as reprinted in Whitford, 74–87.

16. Kirkpatrick Sale provides a concise introduction to some of the central voices on regionalism in *Dwellers in the Land: The Bioregional Vision* (San Francisco: Sierra Club Books, 1985), 137–149.

17. Dowie, 229.

18. The term is used by Stephanie Mills in *Whatever Happened to Ecology* (San Francisco: Sierra Club Books, 1989), as cited in Dowie. The full quote is: "a biological politics so decentralized and wholistic it all but defeats explanation."

19. James McAdoo, "The Role of Transport in Global Pulp and Paper Competitiveness," in Proceedings, Third International Pulp and Paper Symposium: What is Determining International Competitiveness in the Global Pulp and Paper Industry? (Seattle: University of Washington, Center for International Trade in Forest Products, 13–14 September 1994), 248–251. International Institute for Environment and Development (IIED), *Towards a Sustainable Paper Cycle,* final report review draft (London: IIED, 1996), 139–40. For transport costs broken down by grade and regional cost factors (1986 dollars) see Peter J. Ince, *Recycling and Long-Range Timber Outlook: Background Research Report, 1993 RPA Assessment Update, USDA Forest Service* (Madison, WI: U.S. Department of Agriculture, Forest Service, Forest Products Laboratory, 1994), table 54 (transport costs) and tables 48–52 (product costs).

20. International Institute for Environment and Development (IIED), *The Sustainable Paper Cycle, Phase I Review Report* (London: IIED, 1995), annex C, C50-C55.

21. Jane Jacobs writes at length on the topic of import substitution in *Cities and the Wealth of Nations: Principles of Economic Life* (New York: Random House, 1984).

22. For an excellent survey of creative local and regional approaches to economic diversification in timber-dependent regions of the Northwest see Kirk Johnson, "Recycling, Rural Communities, Resource Conservation," *Environment* (November 1993); and, in general, publications from the Northeast Policy Center of the Graduate School of Public Affairs at the University of Washington in Seattle.

23. Bruce Nordman, of the Energy Analysis Program at Lawrence Berkeley Laboratory in California, builds from an interesting analogy between paper efficiency and energy efficiency. Bruce Nordman, "Paper Efficiency: Energy and Beyond," paper presented at Energy Efficient Office Technology 1994, 18 October 1994; and "Celebrating Consumption" (draft paper, August 1995).

24. This discussion of regulatory modes and the TRI is substantially drawn from Robert Gottlieb and Maureen Smith, "The Pollution Control System: Themes and Frameworks," (chapter 1), and Robert Gottlieb, Maureen Smith, Julie Roque, and Pamela Yates, "New Approaches to Toxics: Production Design, Right-to-Know and Definition Debates," (chapter 5) in Robert Gottlieb, ed., *Reducing Toxics: A New Approach to Policy and Industrial Decisionmaking* (Washington, D.C.: Island Press, 1995). Breyer's analysis is in Stephen Breyer, *Regulation and Its Reform* (Cambridge, Mass.: Harvard University Press, 1982).

25. Ironically, in light of the immense impact this information has had since it became available, the corporate voices in the debates that led to its creation seem remarkably subdued. Beyond the fact that excessive protestation could easily have backfired in the immediate aftermath of Bhopal, another reason for the comparatively muted opposition was likely the fact that the debate was conducted within the larger framework of reauthorizing "Superfund" legislation, where unquestionably hard-nosed issues of corporate liability, taxes, and standards were involved. It is also possible that the potential impact of the approach was simply underestimated. See Gottlieb et al., chapter 5.

26. This law, which required manufacturers to present a "clear and reasonable warning" of the presence of state-listed carcinogens and reproductive toxins to which the public might be exposed (including through the use of products), had a demonstrable effect in terms of inducing product reformulation to eliminate hazardous chemical constituents. Randall B. Smith, "California Spurs Reformulated Products," *Wall Street Journal* (1 November 1990); California Environmental Protection Agency, Office of Environmental Health Hazard Assessment, *The Implementation of Proposition 65: A Progress Report* (July, 1992).

27. Forthcoming and suggested expansions of TRI reporting requirements are discussed in numerous EPA publications, including: Proceedings, Toxics Release Inventory (TRI) Data Use Conference: Building TRI and Pollution Prevention Partnerships, Boston, Mass., 5–8 December 1994 (Washington, D.C.: U.S. Environmental Protection Agency, March).

28. Quincy Sugarman, *Breaking the Chemical Dependency: The First Data on Oregon's Industrial Toxics Use* (Portland, Ore.: Oregon State Public Interest Research Group [OSPIRG], 1993), based on 1991 data reported under H.B. 3515, the Oregon Toxics Use Reduction Act of 1989.

29. IIED provides a well-organized and more thorough description of various forms of labeling and certification approaches pertaining to paper. IIED, annex E.

30. Paper Task Force, *Paper Task Force Recommendations for Purchasing and Using Environmentally Preferable Paper* (New York: Environmental Defense Fund, 1995), chapters 4 and 5.

31. From the matrix/poster entitled "Sustainable Forest Products: Opportunity within Crisis" from the John D. and Catherine T. MacArthur Foundation, Chicago, Illinois, 1995.

32. Vance Packard in his 1966 book *The Waste Makers*, as quoted by Jennifer Seymour Whitaker in *Salvaging the Land of Plenty: Garbage and the American*

Dream (New York: William Morrow, 1994), 56. Whitaker provides an excellent discussion of the efforts to shift consumer attitudes from "buying for necessity to the necessity of buying" (54–57, 87–91, and elsewhere).

33. Robert Reich, *The Work of Nations* (New York: Vintage Books, 1992), chapter 5.

34. Daly and Cobb, 1.

Index